CHEMISTRY THE EXPERIENCE

ANN RATCLIFFE

Coordinator, General Chemistry Teaching Laboratories
Oklahoma State University

John Wiley & Sons, Inc.
New York • Chichester • Brisbane • Toronto • Singapore

Notice to users:

Most chemicals are potentially dangerous if they are not handled properly and with respect. Although every reasonable effort was made to test each experiment, and to provide appropriate directions, instructions, and safety data, the author does not assume either responsibility or liability for any mishaps or accidents that may occur in the use of this text as a laboratory manual.

ACQUISITIONS EDITOR / Nedah Rose
MARKETING MANAGER / Catherine Faduska
PRODUCTION SUPERVISOR / Elizabeth Swain
MANUFACTURING MANAGER / Andrea Price

This book was set by the author and printed and bound by Courier Kendallville. The cover was printed by The Lehigh Press, Inc.

On the Cover:

An inexpensive white handkerchief was dipped in a solution of iron(II) sulfate, $FeSO_4$, and then allowed to dry in air. Oxygen, O_2, in the air oxidized the almost colorless iron(II) ion, Fe^{2+}, to yellow iron(III), Fe^{3+}. The procedure was repeated several times to deepen the yellow color. When the handkerchief was dry, it was folded and tied using the techniques illustrated in Appendix D. The tied fabric was dipped in *cochineal*, a dye used by the Aztecs. Cochineal is extracted from *Coccus cacti*, a scale insect that feeds on the cactus *Nopalea coccinellifera* in Mexico. The mordant for the dye, potassium aluminum sulfate, $KAl(SO_4)_2 \cdot 12H_2O$ (known commonly as alum), was produced by the procedure in Inquiry 24.

Recognizing the importance of preserving what has been written, it is a policy of John Wiley & Sons, Inc. to have books of enduring value published in the United States printed on acid-free paper, and we exert our best efforts to that end.

Copyright © 1993 by John Wiley & Sons, Inc.

All rights reserved. Published simultaneously in Canada.

Reproduction or translation of any part of this work beyond that permitted by Sections 107 and 108 of the 1976 United States Copyright Act without the permission of the copyright owner is unlawful. Requests for permission or further information should be addressed to the Permissions Department, John Wiley & Sons, Inc.

Library of Congress Cataloging in Publication Data:

Ratcliffe, Ann.
Chemistry/Ann Ratcliffe.
p. cm.
Includes index.
ISBN 0-471-57707-3 (pbk.)
1. Chemistry. I. Title
QD33.R333 1992
540-dc20 92-27306
CIP

Printed in the United States of America

10 9 8 7 6 5 4 3 2 1

To my parents,

Lanier and Ed,

who encouraged me to inquire

About the Author

Ann Ratcliffe received the A.B. degree in chemistry from Randolph-Macon Woman's College and the Master of Arts in Teaching from Duke University. She taught high school chemistry for eight years, including four years at the American Community School in Beirut, Lebanon. She has been at Oklahoma State University since 1977, where she has been a research associate and a lecturer and, since 1986, has coordinated the general chemistry teaching laboratories. Besides chemical education, her interests include gardening, yoga, electronic music, and the ancient Near East.

Preface

TO THE STUDENT

If you listen carefully to the national news on television you will hear industry and academia expressing a concern about science education in this country. The key phrase is "scientific literacy." Polls and tests show that we have not been doing a very good job of educating the public about issues in the sciences. This laboratory manual is one small effort in the much larger campaign to raise the level of "chemical literacy."

The word "literacy" implies reading and writing, and reading and writing imply *language*. But we don't want to just read and write *about* chemistry, because chemistry is an experimental science. It requires experience, a hands-on approach. Once we have an experience of chemistry we need to be able to communicate it to others. To do that we have to learn the language of chemistry; we have to become literate in chemistry.

The Language of Chemistry

Think for a moment how you learned to speak your native language. Adults spoke to you over and over, and before long you could repeat some of what they were saying. Then you found that you could not only repeat phrases, but you could make sentences of your own, without having someone tell you all the rules of speaking. The rules came much later. This is such a natural way to learn a language that we hardly remember doing it. In fact, it is so natural, that there is a school of language teaching that says that, even for adults, it is better to hear and speak a foreign language first, before having to learn the rules for writing.

But why, you ask, are we talking about foreign language? This is a chemistry course. Precisely! Chemistry, like any subject you learn, has a language all its own. Lab manuals tend to defer to the textbooks, most of which start teaching chemistry in a systematic way that begins with measurement and problem solving, some of the important skills the chemist must develop. This introduction to measurement and basic problem solving is essential, but we don't have to repeat it the first week in lab. There will be ample opportunities in lab to reinforce these concepts. What will help you most with the lecture portion of your course is your having a strong background in the language and patterns of chemistry. Therefore, we begin doing real chemistry in the first Inquiry. By the time your instructor gets to the chapter in the textbook that deals with formula writing and naming, you will already know much of the language of chemistry, and you will be ready to learn *why* we speak and write names and formulas in a certain way.

Safety and the Problem of Waste

The word "chemical" often appears in the press with a negative connotation. We read about hazardous waste dumps leaking deadly chemical poisons into the water table; we worry about dioxin and Agent Orange; we read about the latest chemical spill, or chemical fire, or chemical food contamination. But we sometimes forget that the very basis of our existence on the Earth is chemical. We are chemical; we are made up of the atoms and molecules that are the building blocks of the Universe. Remember that "chemical" is just a word; it is neither negative nor positive. It is what we *do* with chemicals that has ramifications for our future on the Earth. Learning to live and work with chemicals responsibly will be one of our goals.

In this laboratory experience in chemistry there are two major concerns expressed in the design of every Inquiry: 1) that you work with small amounts of low toxicity chemicals in a safe environment, following the prescribed safety rules; and 2) that the waste you produce not burden an already polluted Earth. To these ends, the chemicals chosen are ones that are not toxic to you when used responsibly, and the quantities are so small that the waste they produce can either go safely down the drain (depending on local codes) or can be recycled. In fact, the last Inquiry in the manual is the recycling of the silver that you use in several experiments.

Prelab and Postlab Questions

In each Inquiry there is a section called "Before You Begin." In it you will sometimes find questions that you should answer before you start. Your instructor will probably discuss some of these in introducing the Inquiry, so you will find it useful to have done some background work. Following the Procedure, Observations, and Results, there will be a section called "Follow-up Questions." These will bring together the concepts that you have been studying, developing patterns that will help you extrapolate beyond the Inquiry.

Begin the Adventure

Chemistry can be fun and exciting. It never has to be boring, because you will not be just reading about science, you will be *doing* science. And since it doesn't make much sense for you to simply confirm concepts that have already been taught in lecture, *Chemistry: The Experience* is designed so that you will be discovering for yourself some of the concepts that will be explained more completely in lecture at a later time. For this reason, the exercises are called "Inquiries." They often begin with a question, and, you will discover, often end with more questions! That's the way research works – questions lead to answers which lead to more questions, and so on. That's what makes for excitement and adventure in science. So, let's begin.

TO THE INSTRUCTOR

Chemistry: The Experience is an inquiry-based approach to the teaching of the laboratory portion of the chemistry course for the nonmajor. It can either be integrated into the course, or it can be treated as a *self-integrated* set of Inquiries that use the text as reference, but that do not have to be keyed exactly to the text. By definition, the Inquiries should be done by the students *before* the material is taught in the lecture.

The Self-Integrated Laboratory Experience

Integrating the laboratory into the course is not impossible; most of us do just that when we make certain that the lab portion of the syllabus follows the lecture outline, and when we make an attempt to bring the concepts of the experiments into the lecture. But it is possible to *self-integrate* the lab portion of the course, such that it becomes a mini-course, related to and attached to the lecture part of the course while standing alone as a laboratory experience. Such is *Chemistry: The Experience.* It requires the lecture text as reference, since it does not contain the explanations for the principles that are explored, but it provides the student with the opportunity of examining a concept before it is taught in lecture and of independently seeking the patterns that lead to generalizations.

In *Chemistry: The Experience*, formulas and names are introduced in the first Inquiry, but without explanation. Students are informed that they need not know how formulas are formed; they must simply use them as they are introduced. The method is similar to the oral-aural method of language teaching in which the student begins using the language the first day without knowing all the rules – in other words, as we learn language initially as infants. What the student learns first are the language and methods of chemistry. By the time the instructor reaches formula and equation writing the student has a strong *experiential* background in these areas. Learning to communicate in the language of chemistry then becomes less rote memory and more experience.

The Rationale

As we approach the 21st century, we face issues in science that are staggering in their complexity. Some of these issues – energy, environment, and world food supply, for example – are addressed by the textbooks. In addition, there are two overriding *pedagogical* issues that must be addressed: *scientific literacy* and *chemical responsibility*.

Scientific Literacy

This phrase has become almost a buzzword for the Nineties. It is found not only in professional journals, but is also now heard on the evening news as the U.S. examines its place in a technological world and finds its educational system lacking. *Literacy* implies being able to write a language with ease, and it is the *language* of chemistry that *Chemistry: The Experience* first presents and continues to emphasize as the Inquiries unfold. Every language is composed of patterns that one learns to use to predict meaning and proper usage of new words and phrases. Chemistry is no different, and *Chemistry: The Experience* concentrates on the patterns of nature, helping students understand and apply the scientific method throughout the course, with the goal of instilling the wonder and excitement that comes from recognizing a pattern or regularity and extrapolating to making generalizations that can be tested. Students can discover *for themselves* concepts that were previously unknown to them. That is the excitement of chemistry. If we are going to produce a society educated in science, literate in science, then our methods must be consonant with our goals.

Chemical Responsibility

Chemistry: The Experience makes a serious effort not to reinforce the concept of the "throw-away society." Since the course for which this manual is written generally addresses chemistry in our lives and examines, among other subjects, environmental questions, it seems irresponsible not to address environmental issues of chemical waste. In most lab manuals the students are told to dispose of the waste in "appropriate containers." Although students may be told that the instructor or the storeroom will prepare the waste for disposal in a hazardous waste dump, the student is shielded from that process in the same way that the average individual is protected from actually dealing with his/her refuse beyond the curbside.

Chemistry: The Experience uses only chemicals that are safe for students to handle *responsibly*. The waste is generally nontoxic and, local codes permitting, can go down the drain. Students are taught how to neutralize acids and bases before disposing of them, and thus the act of disposal involves learning chemistry. In the rare instance in which a material should not be disposed of down the drain, it is recycled by the students themselves in an Inquiry that draws attention to hazardous waste issues. An example of this is the fairly frequent use of small amounts of silver nitrate in several Inquiries. The silver waste is collected (usually as the chloride), and in the last Inquiry in the manual the silver waste is divided among the students to be recycled for use the following semester. Students take a certain amount of pride in knowing that they have not only protected the environment from a hazardous compound, but that someone next semester may use what they have produced.

A New Look at Some Traditional Exercises

Although there are some completely new student-tested Inquiries, such as Inquiry 10: "Sunscreeen: Good Cents!" and Inquiry 17: "It's a Gas!", you will recognize many familiar experiments with new twists. Inquiry 14: "Solution Dilution: Answer to Pollution?" uses simple acid-base titration to teach the limitations of procedures and instruments used in analysis. Inquiry 24: "The Ubiquitous Aluminum Can: Dyeing to Recycle" has the familiar aluminum-can-to-alum procedure, but instead of growing an alum crystal, students use the alum as the mordant for a natural dye, quercetin, that they extract from onion skin. They dye pieces of cotton cloth with their onion dye and then learn to tie-dye, experimenting with other natural dyes.

Flexibility

Although the Inquiries are related by common themes and materials, flexibility is built in. Inquiries are linked in Units, but each Inquiry stands alone as well. To avoid equipment limitations, either Inquiry 1 or Inquiry 2 can be used to start the semester, and although 96-well tissue culture plates are specified for use in several of the early Inquiries, spot plates can be used.

Space for Growth

Most laboratory manuals have individual experiments having the same kind of data page in each experiment. *Chemistry: The Experience* assumes that students start the course knowing nothing about the scientific method, and that they can learn to apply it. Therefore, the Observation page is structured in early Inquiries to help students learn how to observe, and then by Inquiry 8 it is a blank page for the student to structure appropriately.

Follow-Up Questions and Projects

Each Inquiry ends with Follow-Up Questions designed to help the student look beyond the Inquiry. The Inquiries in Units 4 and 5 generally have Follow-Up Projects that may be assigned as well. A bibliography in Appendix A starts the student on a literature search.

Detailed Instructor's Manual

The instructor's manual contains a list of necessary chemicals (with amounts for making up solutions) for each Inquiry, where to find special equipment, like the well plates and the conductivity apparatus that the students use in Inquiry 1, where to find diazo paper for Inquiry 10, how to get an unlimited supply of free onion skins, answers to postlab problems, and more.

ACKNOWLEDGMENTS

To my friends and colleagues at OSU I owe a debt of gratitude for their unflagging support: Nancy Gettys and Lucy Pryde graciously read the manuscript and offered valuable suggestions, most of which I have incorporated; Larry Schmitz rekindled my interest in tie-dye, and Karla Rider gave up valuable weekend time to do calculations and to check my work; Corinna Czekaj, Dwaine Eubanks, John Gelder, George Gorin, and Mark Rockley were constant sources of encouragement and good discussions.

My special thanks go to Anna Koester who "had fun" doing the line drawings on her Macintosh IIx using Adobe Illustrator, and to Darrell Berlin, Horacio Mottola, and Don Thompson for help when it was needed.

I am grateful to the reviewers who were willing to bring their experience to bear on this work. Many of their suggestions have become part of the text.

Kenneth Busch
Georgia Institute of Technology

David Lippman
Southwest Texas State University

Henry Dragun
Anne Arundel Community College

James Schreck
University of Northern Colorado

Carolyn Herman
Southwestern College

Carl Snyder
University of Miami

One of the reviewers deserves more than the usual thanks. I met Carolyn Herman last summer, and while she and I were talking about this project she offered to field test the manual at Southwestern College. She used over half of the Inquiries last fall, and her suggestions have been invaluable. I am grateful to her for her enthusiasm and especially for her patience.

The project would not have come to fruition without my editor, Nedah Rose, who saw possibilities in my proposal, and whose phone call welcoming me as a "Wiley author" set me on a path of predawn and posttwilight hours at my Mac Plus that has been both difficult and rewarding. Nedah's humor, enthusiam, and support made it a very real pleasure to complete this work.

Ann Ratcliffe

Safety in the Laboratory

Basic Rules for Laboratory Safety

Although *Chemistry: The Experience* is designed to use low-toxicity compounds and to provide you with a safe and productive lab experience, there are some very basic safety rules that you must follow in any chemistry lab.

SAFETY GOGGLES

Wear **safety goggles** in the laboratory at all times. Although the type of specified eye protection may vary slightly from place to place, it is recommended that you wear splash-proof goggles that fit your face snuggly. These should be worn as soon as the lab starts and should not be removed until you have cleaned up and signed out of the laboratory.

EMERGENCY EQUIPMENT

Be sure you know where the *safety shower*, *eye wash station*, *fire blanket*, and *fire extinguishers* are in your laboratory. Your instructor will show you how to use this equipment.

CLOTHING FOR LAB

Wear clothing that covers your arms, legs, and torso. Shorts and sandals should not be worn in lab, nor should loose, flowing sleeves. An apron may be required. Note any specific requirements in the space provided on the back of this page.

FIRE SAFETY

The Bunsen burner is used often in these Inquiries. Your instructor will show you how to light it and how to use it safely. If you have long hair, tie it back out of the way of the burner, and keep your clothing away from the flame.

SAFETY NOTES

Every Inquiry has a section before the Procedure called "Safety Notes." This section points out specific hazards in the Inquiry. Read the Safety Notes carefully before you go to lab, and if you have any questions about safe procedures, discuss them with your instructor.

Personal Responsibility

In the Safety Notes before each Procedure, following the specific information about FLAMMABLE or CORROSIVE materials used in that Inquiry, is the following statement:

> All chemicals must be handled carefully and treated with respect. The solids and liquids used in this Inquiry are safe for you to use responsibly.

The key word here is, of course, "responsibly." This includes not only your responsibility to yourself, to keep yourself from harm, but also your responsibility to your classmates in lab. It also includes responsibility to the environment. Although most of the chemicals you will use can be disposed of down the drain or in the trash, some may be recycled, and there may be specific procedures given in the Inquiry or by your instructor for disposing of the waste from the Inquiry. Follow these instructions carefully. Remember that each of us makes a difference in the quality of life on our planet.

Additional Safety Instructions

The chemistry department may have additional rules that you are expected to follow. Note them here.

Special Equipment

Some of the equipment you will use in this course will be standard laboratory equipment, such as the test tubes, beakers, and flasks that you probably used in science courses in secondary school. However, since one of our concerns in this laboratory manual is responsible handling of waste, most of the Inquiries are done with reduced amounts of chemicals. This micro-method of experimenting uses some special equipment, which is illustrated below. Your instructor will indicate when other equipment is substituted for these in the Inquiries.

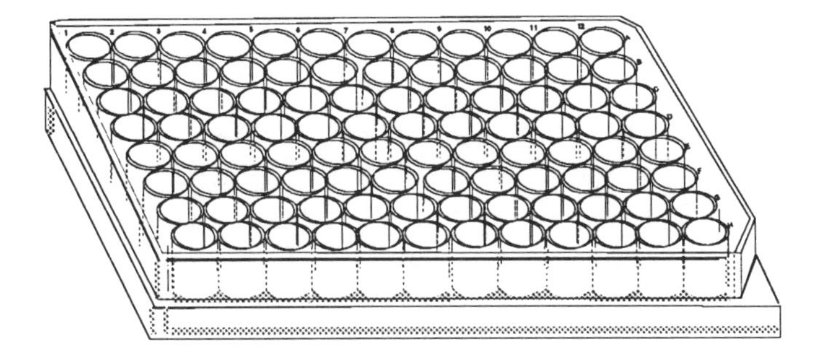

Fig. 1: 96-well plate

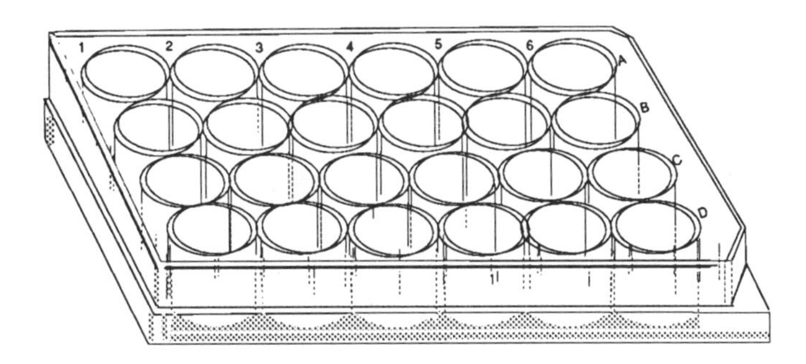

Fig. 2: 24-well plate

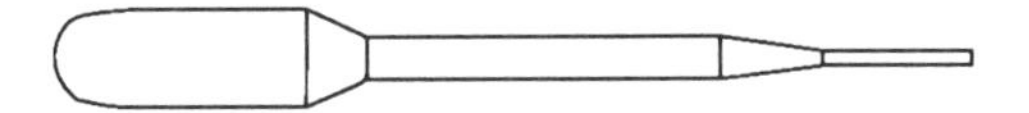

Fig. 3: Plastic (disposable) pipet

Contents

Unit 5: Chemistry for the Future 183

Unit 1

What Does a Chemist Do?

The Scientific Method

The cornerstone of the scientific method is *observation*. The chemist *observes* a phenomenon. Take, for example, the observation that table salt dissolves in water. The chemist might ask several questions: "Why does salt dissolve in water?" "Do other substances dissolve in water?" "Does it matter whether the water is cold or hot?" "Does salt dissolve in other liquids besides water?"

The chemist might then design a simple experiment that will yield many observations, and he or she can look for *patterns* or *regularities* in the observations. For example, a simple experiment that would answer one of the questions above would be to try to dissolve other solids besides table salt in water. If some substances dissolve in water and others do not, then there may be some similarity among those that *do* dissolve in water that can lead to a *generalization* about substances which dissolve in water. "All substances which are . . . . dissolve in water."

The next step might be to formulate a *hypothesis* that attempts to explain why the generalization is true. "The *reason* that all substances which are . . . dissolve in water is . . . ." Now the chemist may make *predictions* about other similar substances, and this will lead to further experimentation, more observations, and doubtless to new questions. In this Unit you will apply the scientific method and look at the results.

Using the Language of Chemistry

When you make observations, you must be able to communicate them to others. So the first thing you will need to be able to do in this Unit is to read chemical formulas correctly. Formulas are not only the chemist's shorthand notation, but, as you will discover later in the course, formulas impart a great deal of information. For the first Unit it is sufficient for you to be able to pronounce them. For example, sodium chloride is written "NaCl" and is pronounced one letter at a time, or "n-a-c-l." When you *say* the formula, it is not important to say which letters are capitals and which are lower case. However, when you *write* a formula, you must pay special attention to the capital and lower case letters. Remember, this is a language!

So, try these: Sulfuric acid, H_2SO_4, is "H-2-S-O-4." Calcium carbonate, $CaCO_3$, is "C-a-C-O-3." Now these: ammonium chloride, NH_4Cl; magnesium bromide, $MgBr_2$. Good. One final point, and a very important one: There is a rule you must follow in Unit 1. It is so important that it is called simply, **THE RULE.** It is this: EVERY TIME YOU WRITE THE NAME OF A COMPOUND, YOU MUST WRITE ITS FORMULA AS WELL. That's all there is to it. So if the answer to a question is "sodium chloride," you must write "sodium chloride, NaCl." If your observation concerns aluminum bromide, you must write "aluminum bromide, $AlBr_3$." All formulas are given in the Inquiries, so you do not have to memorize any formulas now. So, go on to the first Inquiry, and remember THE RULE!

INQUIRY 1

What Do You Observe?

Chemists are fascinated by the physical world and the changes it undergoes. To organize their thinking they use a systematic procedure known as the *scientific method* which begins with *observations*. In this Inquiry you will test some solids and liquids with a test apparatus, *observe* what happens, and seek *patterns* in your observations.

Equipment you will need

beaker (for Cleanup)
test apparatus
96-well tissue culture plate

Other chemicals

Solids

calcium chloride, $CaCl_2$
sand, SiO_2
sodium chloride, NaCl (table salt)
sucrose, $C_{12}H_{22}O_{11}$ (table sugar)

Chemicals you will use

Liquids and solutions, 0.1*M*, in dropper bottles (see Before You Begin, below)

acetic acid, CH_3COOH
ammonia, $NH_3(aq)$
ammonium chloride, NH_4Cl
calcium chloride, $CaCl_2$
copper(II) sulfate, $CuSO_4$
ethanol, C_2H_5OH
glucose, $C_6H_{12}O_6$
hydrochloric acid, HCl
potassium chloride, KCl
potassium hydroxide, KOH
propanol, C_3H_7OH
sodium carbonate, Na_2CO_3
sodium chloride, NaCl
sodium hydroxide, NaOH
sucrose, $C_{12}H_{22}O_{11}$
sulfuric acid, H_2SO_4
test solution
urea, NH_2CONH_2
water, distilled, H_2O

BEFORE YOU BEGIN

1. Pronounce the formulas of the compounds above. If you are unsure about how to do this, turn back one page and review the introduction to Unit I.

2. The "*M*" in "0.1*M*" is read "molar." Look up *molarity* in your text. Write its definition and underline any words which need further definition. You will learn more about molarity later in the course.

3. Your instructor will demonstrate the use of the test apparatus with the 96-well plate. Take notes here.

SAFETY NOTES

1. Wear approved safety goggles at all times in the lab.

2. All chemicals must be handled carefully and treated with respect. The solids and solutions used in this Inquiry are safe for you to use responsibly.

PROCEDURE

1. Set the 96-well plate before you on the lab bench and examine it carefully. In several Inquiries this semester it will serve as 96 micro test tubes. Note that there are numbers across the top of the plate and letters down the sides. Each well can be identified by a row letter and a column number: A1, B3, etc. This will be important in keeping track of the chemicals you have in the various wells. Tear out the Inquiry 1 Observations page, place it next to your 96-well plate, and record your observations in the appropriate space as you do each step of the Procedure.

2. Put 3 or 4 drops of the test solution in well A1 and introduce the wire leads of the test apparatus into the liquid. The light should glow brightly. If it does not, check with your instructor. Record your observations in well A1 on the Observations page.

3. Rinse the wire leads with distilled water and dry them. Test each liquid in the Chemicals list on p. 3 by placing 3 or 4 drops of the liquid in a clean well and testing with the test apparatus as in step 2. Try to keep the leads the same distance apart in each well. On the Observations page, write the name and formula of each compound and record the effect of each liquid on the test apparatus. Now carefully place a small amount of each solid (just enough to cover the bottom of the well) in *dry* wells and test them, recording your observations. Rinse and dry the wire leads after each use.

CLEANUP

(All Cleanup suggestions are subject to local laws governing waste from laboratories. The following are suggestions only and may be changed by your instructor. Space has been provided for additional instructions.)

1. The materials you have used are not harmful to the environment in small quantities. Rinse the contents of the 96-well plate into a beaker and allow any undissolved solid to settle. Pour the liquids down the drain, and wipe the remaining solid out with a paper towel. Throw the towel in the trash.

2. Run plenty of water over the 96-well plate and shake it dry before returning it to the storeroom or putting it away. Rinse the leads of the test apparatus, and dry them. Rinse and dry the beaker.

3. Return all chemicals to their proper places in the lab, and wash and dry your lab bench. Wash your hands before leaving the lab.

4. Additional instructions:

Name ______________________________ Sec________

INQUIRY 1: OBSERVATIONS

Be sure to write the name and formula of each substance tested in the cell below, and record your observations there as well. (cont'd next page)

	1	2	3	4	5	6
A						
B						
C						
D						
E						
F						
G						
H						

Name ______________________ Sec________

INQUIRY 1: OBSERVATIONS (cont'd)

Be sure to write the name and formula of each substance tested in the cell below, and record your observations there as well.

	7	8	9	10	11	12
A						
B						
C						
D						
E						
F						
G						
H						

Name ____________________ Sec ____________

INQUIRY 1: FOLLOW-UP QUESTIONS

These are to be done in the laboratory after the Inquiry. You are encouraged to discuss these with your lab partner and your lab instructor.

1. Carefully examine your worksheet and the names and formulas of the substances you tested. What *patterns* or *regularities* do you see? Note three of them.

   Example: "The substances in this Inquiry which contain the name 'acid' caused the test apparatus to light up."

   a.

   b.

   c.

2. Once a pattern is observed, you might make a *generalization*. For example, a generalization you could make from your observations is that "all acids cause the test apparatus to light up." To test this generalization it would be necessary to test other acids to see if they indeed fit your generalization. Use the index of your textbook to find "acids," and write the names and formulas of five other acids.

   a.

   b.

   c.

   d.

   e.

3. One of the most important tools a chemist has is the periodic table of the elements. Locate the periodic table and the accompanying alphabetical list of elements with atomic weights inside the front and rear covers of your lab manual. There are several numbers in each square of the periodic table, but one of the numbers increases by 1 as you move across any row of the table. This is the *atomic number*. You will study the atomic number in detail in your lecture section later in the course, but for now we will use it to help us identify the elements we have used in this Inquiry and to look for patterns.

   a. In the list of elements, find *sodium, Na*. What is its *atomic number*? ______

   b. Find *potassium, K*. What is its atomic number?______

Name ______________________________________Sec______________

INQUIRY 1: FOLLOW-UP QUESTIONS (cont'd)

c. The periodic table is laid out in rows and columns. The columns are called *families* or *groups*. Look at the top of the column in which sodium, Na, and potassium, K, are found. What group number is there?

____

d. In what group are magnesium, Mg, and calcium, Ca, found? __________

e. In what group do you find chlorine, Cl? __________

f. Locate carbon, C, hydrogen, H, and oxygen, O. Write the group number of each one.

C __________

H __________

O __________

4. If solutions of sodium chloride, NaCl, and potassium chloride, KCl, in water both cause the test apparatus to light up, what do you think *might* be true of solutions of lithium chloride, LiCl, rubidium chloride, RbCl, and cesium chloride, CsCl? (Look at the periodic table and the element list.)

5. The next step in the scientific method after making generalizations is the formation of a *hypothesis*. The hypothesis is a possible explanation about what might be occurring when the light goes on. Write a hypothesis that *might* explain the observation that the light responds to certain substances and not to others.

6. Each Inquiry that you do should raise questions that are not answered by the Inquiry. Write two questions that this Inquiry raises about which you are curious.

   a.

   b.

INQUIRY 2

Does It Dissolve in Water?

You know that some solids dissolve in water and some do not. You have dissolved sugar in iced tea and salt in water, and you have seen a stream flowing over rocks that do *not* dissolve appreciably. In order to ultimately understand how solution in water occurs, it is first necessary to test many solids, make *observations*, and then seek *patterns* or *regularities* in the observations.

Equipment you will need

microstirrer
test tubes, 75 mm (2)
96-well plate

Chemicals you will use

Solids

aluminum sulfate, $Al_2(SO_4)_3$
ammonium chloride, NH_4Cl
ammonium sulfate, $(NH_4)_2SO_4$
benzoic acid, C_6H_5COOH
calcium carbonate, $CaCO_3$
calcium chloride, $CaCl_2$
calcium nitrate, $Ca(NO_3)_2$
calcium sulfate, $CaSO_4$
copper(II) carbonate, $CuCO_3$
copper(II) chloride, $CuCl_2$
copper(II) nitrate, $Cu(NO_3)_2$
copper(II) sulfate pentahydrate, $CuSO_4 \cdot 5H_2O$
glucose, $C_6H_{12}O_6$
potassium bromide, KBr
potassium carbonate, K_2CO_3
potassium chloride, KCl
potassium nitrate, KNO_3
potassium sulfate, K_2SO_4
sand, SiO_2
silver chloride, AgCl
silver nitrate, $AgNO_3$
sodium bromide, NaBr
sodium carbonate, Na_2CO_3
sodium chloride, NaCl
sodium nitrate, $NaNO_3$
sodium sulfate, Na_2SO_4
sucrose, $C_{12}H_{22}O_{11}$
urea, NH_2CONH_2

water, distilled, H_2O (in dropper bottle)

BEFORE YOU BEGIN

1. You will find that it is sometimes easier to see the solid you are testing if the well plate is sitting on a black bench. This is particularly true if the solid is white. If the solid is dark, or the solution it makes is colored, you may find it easier to see what is occurring if the well plate is sitting on white paper. Have a piece of paper handy to slip under the plate if you find it necessary.

2. Review the Unit I introduction, pp. 1-2. A new type of formula appears in this Inquiry. Remember that to pronounce the formula for sodium chloride, you just read the letters "N-a-C-l." Or, for sodium carbonate, Na_2CO_3, you would say "N-a-2-C-O-3." But what if there are parentheses in the formula? For example, aluminum sulfate has the formula $Al_2(SO_4)_3$. One way to read this formula is "A-l-2-S-O-4-taken 3 times."

   a. How would you pronounce calcium phosphate, $Ca_3(PO_4)_2$?

   ____

   (For "2 times," you can say "twice.")

   b. What about ammonium sulfate, $(NH_4)_2SO_4$?

   ____

3. The chemicals you use will be in small bottles with caps or stoppers. It is *imperative* that a cap or stopper be replaced *immediately* after you remove the amount of chemical you need. There are two good reasons to do this:

   a. Open bottles with their tops left lying on the table invite contamination, as the wrong top may be put on a bottle

   b. Some chemicals are changed by moisture in the air, so the bottles must be kept tightly capped.

SAFETY NOTES

1. Wear approved safety goggles at all times in the lab.

2. All chemicals must be handled carefully and treated with respect. The solids used in this Inquiry are safe for you to use responsibly.

3. Compounds containing silver, Ag, will cause a stain on skin and clothing. Always wash your hands immediately after working with silver compounds.

4. Two compounds used in this Inquiry, silver chloride, AgCl, and silver nitrate, $AgNO_3$, cannot be disposed of down the drain. Silver compounds are toxic if ingested. They are also quite expensive, so your instructor will show you where to empty the small test tubes containing these substances. All the silver compounds used this semester will be collected to be recycled. Your instructor may assign Inquiry 25 in which you recycle the silver you and other students have used.

5. Wash your hands before you leave the lab.

PROCEDURE

1. Tear out the Inquiry 2 Observations page and lay it beside your 96-well plate. Because solids are somewhat difficult to introduce into the wells without spilling, you may find it helpful to use every other column of the well plate.

2. Since silver chloride, AgCl, must be tested separately, instead of using a well in the well plate use a very small test tube. Put a sample of the chemical *no larger than the head of a pin* in the test tube, and add a few drops of water. Shake the test tube, and record in the first space (A1) in your Observations page whether the silver chloride, AgCl, dissolves. (Remember to write names and formulas for each observation.) Do not empty this test tube into the sink. See below under Cleanup for instructions.

3. Repeat using one or two crystals of silver nitrate, $AgNO_3$, in a fresh test tube, and record your observations in space A2 of the Observations page. Again, check Cleanup for instructions for disposal.

A well

A well with a sample

4. For the rest of the chemicals in the list you may use the well plate. Starting with aluminum sulfate, $Al_2(SO_4)_3$, place a sample the size of *3 grains of sand* in well A3. The sample must be very small, because a larger sample might not dissolve completely, making it difficult to determine if any solution had taken place. If the sample is made up of tiny crystals, one or two crystals will suffice.

5. Using a dropper, add a few drops of water. Stir with the microstirrer, and record your observations on the Inquiry Observations page. Some solids dissolve slowly, so you may want to reserve judgment until you have tested all the solids. Your observations should include everything you observe about the solid and its behavior in water. Be sure to rinse the microstirrer with a few drops of water and dry it after each use.

6. Continue testing and recording your observations until you have tested each solid. If you are unsure of the results in any well, repeat the test.

CLEANUP

(All Cleanup suggestions are subject to local rules governing waste from laboratories. The following are suggestions only and may be changed by your instructor. Space has been provided for additional instructions.)

1. The materials you have used, except silver chloride, AgCl, and silver nitrate, $AgNO_3$, are not harmful to the environment in small quantities and may be disposed of by rinsing them down the drain. Empty the test tubes containing the silver chloride, AgCl, and silver nitrate, $AgNO_3$, into the waste container. Rinse the tubes with a few drops of water, and add the rinsings to the waste container.

2. Empty your well plate into the sink. Run plenty of water over it, and shake it dry before returning it to the storeroom or putting it away.

3. Return all chemicals to their proper places in the lab, and wash and dry your lab bench. Wash your hands before you leave the lab.

4. Additional instructions:

Name ________________________ Sec________

INQUIRY 2: OBSERVATIONS

Be sure to write the name and formula of each substance tested in the cell below, and record your observations there as well. (cont'd next page)

	1	2	3	4	5	6
A						
B						
C						
D						
E						
F						
G						
H						

Name ______________________ Sec______

INQUIRY 2: OBSERVATIONS (cont'd)

Be sure to write the name and formula of each substance tested in the cell below, and record your observations there as well.

	7	8	9	10	11	12
A						
B						
C						
D						
E						
F						
G						
H						

Name ______________________________ Sec ____________

INQUIRY 2: FOLLOW-UP QUESTIONS

These are to be done in the laboratory after the Inquiry. You are encouraged to discuss these with your lab partner and your lab instructor.

1. Another way to ask, "Does it dissolve in water?" is to ask, "Is it soluble in water?" Write three *patterns* or *regularities* that you find in your observations concerning the solubility in water of the compounds in this Inquiry.

   a.

   b.

   c.

2. One of the most important tools a chemist has is the periodic table of the elements. Locate the periodic table and the accompanying alphabetical list of elements with atomic weights inside the front cover of your lab manual. There are several numbers in each square of the periodic table, but one of the numbers increases by 1 as you move across any row of the table. This is the *atomic number*. You will study the atomic number in detail in your lecture section later in the course, but for now we will use it to help us identify the elements we have used in this Inquiry and to look for patterns.

   a. In the list of the elements, find aluminum, Al. What is its atomic number? ________

   b. Find calcium, Ca. What is its atomic number? ________

   c. Locate boron, B, helium, He, and nitrogen, N. Write the atomic number of each one.

      B ________

      He ________

      N ________

   d. The periodic table is laid out in rows and columns. The columns are called *families* or *groups*. Look at the top of the column in which calcium, Ca, is found. What group number is there? ________

   e. In what group do you find each of the following?

      aluminum, Al ________

      copper, Cu ________

      bromine, Br ________

      chlorine, Cl ________

Name ______________________________ Sec ____________

INQUIRY 2: FOLLOW-UP QUESTIONS (cont'd)

3. a. List the names (and the formulas!—remember THE RULE!) of all the compounds in this Inquiry in which the first element in the name is in Group I of the periodic table.

   b. Based on your observations in this Inquiry, what generalization can you now make about the solubility of the compounds of Group I elements?

4. What generalization can you make about compounds containing copper, Cu?

5. If you were given a mixture of solid sodium chloride, NaCl, and solid silver chloride, AgCl, how might you separate them so that when you were finished, the sodium chloride, NaCl, and the silver chloride, AgCl, would be in different containers?

6. You should have many questions that have been inspired by this Inquiry. Write 2 of them here.

   a.

   b.

INQUIRY 3

Some Pennies for Your Thoughts

So far we have done chemistry without using that most important tool: *measurement.* Although there may seem to be no limit to the chemistry we can do without measurement, our horizons are greatly expanded by being able to *quantify* what we do.

Equipment you will need

balance

Chemicals you will use

no laboratory chemicals necessary

Other items

magnifying glass
metric ruler
pennies (10, from the storeroom)

BEFORE YOU BEGIN

1. In this Inquiry you will use the concept of *density.* Look in your textbook and find the definition of *density.* Write it here.

2. Now write the definition of density as a *mathematical formula.*

3. To determine the density of a penny, then, it will be necessary to find its mass, m, and its volume, V. Finding mass is easy; you simply weigh the penny. To find volume you can measure the penny's dimensions and calculate the volume. To do this you have to know two things:

   a. What kind of geometrical shape does a penny have?

   b. What is the formula for the *volume* of that shape?

   $V =$

4. Draw a sketch of a penny in the margin, and label the dimensions you will have to measure.

5. Define "mint" as it relates to coins.

SAFETY NOTES

1. Wear approved safety goggles at all times in the lab.

PROCEDURE

Part I: Mass

1. Obtain 10 pennies from your instructor or from the storeroom.

2. Tear out the Inquiry 3 Observations page and take it with you to the balance. Weigh each penny and record on the Observations page its mass and the year it was minted. If you keep the pennies stacked in the order in which you weigh them, it will be easier to make and record the observations in Part II.

Part II: Volume

1. Using a metric ruler and a magnifying glass, measure the diameter, *d*, and thickness (or height), *h*, of each penny. Record these observations in the table on the observation page. Keep them in the same order for step 3 below.

2. Place your observations of mass, diameter, and height on the chalkboard with those of the other students. Look at all the masses and dates. Do you see a pattern? Record your observations. Does the same pattern exist for diameter and height?

3. There is another simple way to determine the volume of a penny. Talk it over with your lab partner, and decide how you will proceed. Turn to p. 22, and write a short procedure. Have it approved by your instructor before you start. Record your observations on the same page.

CLEANUP

(All Cleanup suggestions are subject to local laws governing waste from laboratories. The following are suggestions only and may be changed by your instructor. Space has been provided for additional instructions.)

1. When you have finished, put away all equipment, clean your bench, and wash your hands before leaving the laboratory.

2. Additional instructions:

Name ______________________________ Sec__________

INQUIRY 3: OBSERVATIONS

1. Fill in the table with your observations from Procedures I,2 and II,1.

Penny	Year	Mass, g	Diam., mm	Height, mm
1				
2				
3				
4				
5				
6				
7				
8				
9				
10				

2. Look at the collected observations of mass, diameter, and height on the chalkboard. What pattern(s) do you observe?

Name ________________________________ Sec ____________

INQUIRY 3: CALCULATIONS AND RESULTS

1. Calculate volume, V, and density, D, writing in the space below the formulas you found (p. 17). How does the radius of a circle, r, compare to the diameter?

Penny	Year	radius, r	$V =$	$D =$
1				
2				
3				
4				
5				
6				
7				
8				
9				
10				

2. What regularity exists in the densities of pennies?

3. What are the *units* of the densities you have calculated?

Name ____ Sec______________

INQUIRY 3: FOLLOW-UP QUESTIONS

These are to be done in the laboratory after the Inquiry. You are encouraged to discuss these with your lab partner and your lab instructor.

1. How do you suppose the federal government was able to decrease the mass of a penny while keeping its volume the same?

2. If a "new" penny weighs 2.50 g and is composed of 97.6% zinc, Zn, and 2.40% copper, Cu, what is the mass of zinc in the penny? The mass of copper?

3. Why was it necessary to combine your observations of masses and mint years with those of other students?

4. What generalization can you now make about pennies?

5. If you were to do research on pennies, what are two questions that you would like to answer?

   a.

   b.

Name ____ Sec ______________

Alternate Procedure to Determine Volume of a Penny:

INQUIRY 4

Taking Apart a Mixture: Microscale

Chemists are often faced with the challenge of separating one substance from another. For example, iron ore might contain a mixture of magnetite, Fe_3O_4 (an oxide of iron), and silica, SiO_2. The chemist's job would be to obtain pure iron, Fe, from this mixture. Some steps in the process of refining the ore involve chemical reactions, and some involve physical separations. Separating the components of iron ore is a complex operation, but we can do a simpler separation to illustrate the stepwise thinking a chemist must use in designing any experiment. In this Inquiry you will use micro techniques to separate a mixture of two compounds that you have examined earlier, copper(II) sulfate pentahydrate, $CuSO_4{\cdot}5H_2O$, and urea, NH_2CONH_2, based on their solubilities in two different solvents, water, H_2O, and ethanol, C_2H_5OH. Then you will learn a technique for confirming the identity of a compound (and for indicating its purity).

Equipment you will need

Bunsen burner
capillary tubes, 1.8 mm
droppers
microstirrer
rubber band
thermometer, -20–200°C
Thiele tube or large test tube
watch glasses (2)
96-well plate

Optional

oven, heat lamps, or hot plates
wax for Thiele tube

Chemicals you will use

Solids

copper(II) sulfate pentahydrate, $CuSO_4{\cdot}5H_2O$
urea, NH_2CONH_2

Liquids

ethanol, C_2H_5OH
water, distilled, H_2O

BEFORE YOU BEGIN

1. Several new terms are used in this Inquiry. Before lab, use your textbook and write definitions of the following.

a. solution:

b. solvent:

c. soluble:

d. mixture:

e. compound:

2. In this Inquiry, you will use two solids, copper(II) sulfate pentahydrate, $CuSO_4 \cdot 5H_2O$, and urea, NH_2CONH_2. At first glance, there is one obvious property of copper(II) sulfate pentahydrate, $CuSO_4 \cdot 5H_2O$, that distinguishes it from urea, NH_2CONH_2. If you don't remember what that is, look at the two solids when you get to lab, and write that property below:

____

3. Your instructor will demonstrate how to set up a Theile tube or a large test tube containing wax or oil for determining the melting point of urea, NH_2CONH_2. She or he will also demonstrate how to close one end of a capillary tube and how to put about 0.5 cm of urea in the tube. Take notes here.

SAFETY NOTES

1. Wear approved safety goggles at all times in the lab.

2. Ethanol, C_2H_5OH, is FLAMMABLE. Keep it away from the Bunsen burner flame used to close the capillary tube in Part II of the Procedure.

3. All chemicals must be handled carefully and treated with respect. The solids and liquids used in this Inquiry are safe for you to use responsibly.

4. Wash your hands before you leave the lab.

PROCEDURE

Part I: Separation

In this part it will be necessary for you to do some testing of your solvent and solutes first and to answer some questions which will help you construct a procedure for the separation.

1. In Inquiry 2 you tested the solubility of copper(II) sulfate pentahydrate, $CuSO_4{\cdot}5H_2O$, and urea, NH_2CONH_2, in water. If you have those observations with you, write them on the Observations page for this Inquiry. If you don't have them, use your 96-well plate and check their solubilities in water. Remember to use a tiny amount, *about the size of a pin head.* Record your all your observations on the Inquiry 4 Observations page.

2. Ethanol, C_2H_5OH, is a solvent with properties different from those of water. Place a tiny sample of each of the two solids in separate wells, add a few drops of ethanol, C_2H_5OH, and stir. (You may need to break the crystals with the microstirrer.) Record your observations.

3. In another well, mix a tiny sample of each of the two solids. Now look carefully at your observations. Which solvent dissolved both solids?

____

Which solvent dissolved only one of the solids?

____

So which solvent will you use *first* to remove one of the solids?

____

Check your answers with your instructor before proceeding.

4. Add a few drops of the chosen solvent to the mixture of solids and stir for a minute or two. What happens? Record your observations.

5. Let the remaining solid settle, and then, using a dropper, remove the liquid portion, leaving the solid behind. Place the liquid on a watch glass. Add a few additional drops *of the same solvent* to the solid remaining in the well, and stir again. Let the solid settle, and, using the dropper, suction again to remove any solution, adding this solution to the other portion on the watch glass. Why is it advisable to add solvent a second time?

6. Crystals remain in the well. To remove them, first add several drops of the second solvent, and stir. After the crystals dissolve, remove them with a dropper and place the solution on another clean watch glass. To be sure you removed all the solid, add a few more drops of solvent to the well, stir, and add that solution to the watch glass.

7. The two solvents will evaporate in air if left in your locker for several hours (or for a week), but a more practical way to evaporate the solvents quickly is to place the watch glasses under

a heat lamp or in a warm (~80°C) oven or on a warm hot plate. Your instructor will tell you what to do. After the solvents have evaporated, observe the results.

Part II: Melting Point of Urea

1. Set up the melting point apparatus as you were shown. Obtain two capillary tubes, and close one end of each in a Bunsen burner flame.

2. Obtain a small sample (size of a match head) of urea, NH_2CONH_2, on a clean watch glass. Insert a few crystals of the solid into the open end of one of the capillary tubes by placing the tube with open end down into the solid several times, forcing crystals into the tube. Tap the closed end gently on the bench to tamp down the crystals. Repeat until about 0.5 cm of crystals are packed in the tube.

3. Set up the second capillary tube as above using the urea, NH_2CONH_2, that you recovered from the mixture. Attach both capillaries to the thermometer with a rubber band as your instructor demonstrated.

4. Begin warming the wax or oil in the melting point apparatus. Since urea, NH_2CONH_2, has a melting point above 100°C, you may heat rapidly to 100°C and then slow the heating to about 2°C per minute. Watch the two capillaries and write down the temperatures at which melting takes place.

CLEANUP

(All Cleanup suggestions are subject to local laws governing waste from laboratories. The following are suggestions only and may be changed by your instructor. Space has been provided for additional instructions.)

1. The materials you have used are not harmful to the environment in small quantities and may be disposed of by rinsing them down the drain.

2. Run plenty of water over your 96-well plate and shake it dry before returning it to the storeroom or putting it away.

3. Place the used capillaries in the container for glass waste.

4. Return all chemicals to their proper places in the lab, and wash and dry your lab bench. Wash your hands before you leave.

5. Additional instructions:

Name ____Sec______________

INQUIRY 4: OBSERVATIONS

1. Are the two solids, copper(II) sulfate pentahydrate, $CuSO_4 \cdot 5H_2O$, and urea, NH_2CONH_2, soluble in water, H_2O?

   copper(II) sulfate pentahydrate, $CuSO_4 \cdot 5H_2O$ ____________________

   urea, NH_2CONH_2 ____________________

2. Are the two solids, copper(II) sulfate pentahydrate, $CuSO_4 \cdot 5H_2O$, and urea, NH_2CONH_2, soluble in ethanol, C_2H_5OH?

   copper(II) sulfate pentahydrate, $CuSO_4 \cdot 5H_2O$ ____________________

   urea, NH_2CONH_2 ____________________

3. What happens when you add ethanol to the mixture of the two solids?

4. Describe the appearance of the materials remaining on the watch glasses after the solvents have evaporated.

   a.

   b.

5. Have you effectively separated the two solids?

6. Melting temperatures: pure urea ________°C

   urea from mixture ________°C

7. Write your observations of the melting points on the board. What can you conclude from the melting point determinations?

Name ______________________________ Sec__________

INQUIRY 4: FOLLOW-UP QUESTIONS

These are to be done in the laboratory after the Inquiry. You are encouraged to discuss these with your lab partner and your lab instructor.

1. Chemists are particularly fond of communicating information graphically. (It is sometimes true that a picture is worth a thousand words!) Take, for example, the problem of separating a mixture of table salt and sand. By now it is obvious to you that salt dissolves in water, and your previous experience would tell you that sand is not very soluble in water. So, to separate them, one could simply add water, thereby dissolving the salt, decant (pour off) the salt water solution into another container, and evaporate the water from the salt. The sand is left behind in the first container. To show this graphically, the chemist uses a "flowchart." Flowcharts are also common in computer programming and in business. The flowchart for the separation of salt and sand would look like this.

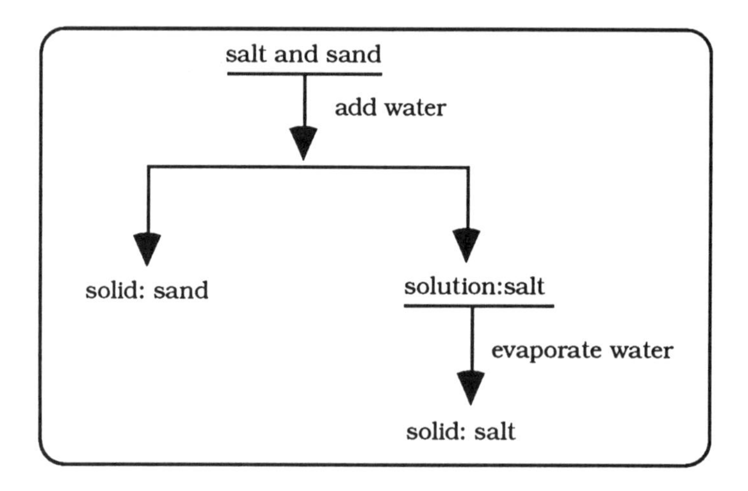

Draw a flowchart that represents the separation of the two solids in this Inquiry: copper(II) sulfate pentahydrate, $CuSO_4{\cdot}5H_2O$, and urea, NH_2CONH_2. Use p. 30 if you need more space.

copper(II) sulfate pentahydrate, $CuSO_4{\cdot}5H_2O$, and urea, NH_2CONH_2

Name ______________________________ Sec__________

INQUIRY 4: FOLLOW-UP QUESTIONS (cont'd)

2. Assume you have run a series of solubility tests of solids in water. The results you obtain are as follows:

   a. Sodium chloride, $NaCl$, is soluble in hot and cold water.
   b. Benzoic acid, C_6H_5COOH, is not soluble in cold water, but is soluble in hot water.
   c. Silver bromide, $AgBr$, is not soluble in either hot or cold water.

   If you were given a mixture of the three solids, how would you separate them? Answer with a flowchart.

3. Write two questions that are inspired by this Inquiry.

INQUIRY 5

Taking Apart a Mixture: A Quantitative Approach

In Inquiry 4 two solids were separated by physical means. Think how much more we would know about the effectiveness of the separation technique we used if we knew *how much* of each chemical we had started with and how much we recovered. To make the problem more interesting, we'll add sand, SiO_2, to the mixture of copper(II) sulfate pentahydrate, $CuSO_4 \cdot 5H_2O$, and urea, NH_2CONH_2. And to make it even more challenging, there is no Procedure given; you must design your own procedure in Before You Begin, below, and bring it to lab with you. The same solvents, distilled water, H_2O, and ethanol, C_2H_5OH, will be used.

Equipment you will need

beakers (several) small
Bunsen burner
droppers
evaporating dishes
graduated cylinder(s)
hot plate
oven (optional)
test tubes, 100 mm
wash bottle

Chemicals you will use

copper(II) sulfate pentahydrate, $CuSO_4 \cdot 5H_2O$
ethanol, C_2H_5OH
sand, SiO_2
urea, NH_2CONH_2
water, distilled, H_2O

BEFORE YOU BEGIN

1. In Inquiry 4 you separated a micromixture of copper(II) sulfate pentahydrate, $CuSO_4 \cdot 5H_2O$, and urea, NH_2CONH_2, and you drew a flowchart to represent the separation. Draw that flowchart here.

2. In order to place the sand, SiO_2, in your flowchart, what property of sand must you know?

3. Based on prior experience (or on your observations of sand in either Inquiry 1 or 2) you can make an *assumption* (an educated guess) about the position that sand will occupy in your flowchart. You can confirm your assumption at the beginning of the procedure below. Draw your projected flowchart here.

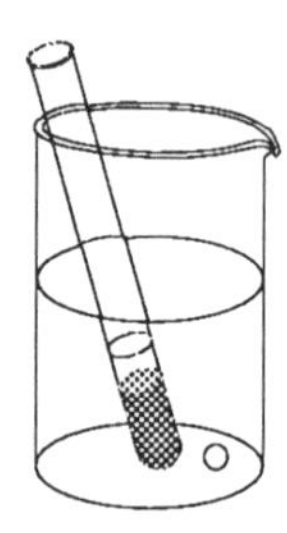

4. To make your flowchart into a procedure that you can follow, you must add more specific instructions. For example, if on one of the arrows you have written "Add ethanol, C_2H_5OH," then you will need to decide *how much* ethanol to add. When one dissolves a solid to remove it from a mixture it is always a good idea to add at least 2 portions of solvent. So you might decide to add 3 mL of solvent, stir to dissolve the solid, and decant, and then add an additional 2 mL of the same solvent, stir, and decant to be sure that all the solid has dissolved. You may also find that heating the test tube containing the mixture and solvent in a water bath (see illustration) will speed the solution process. You may write your procedure on p. 38.

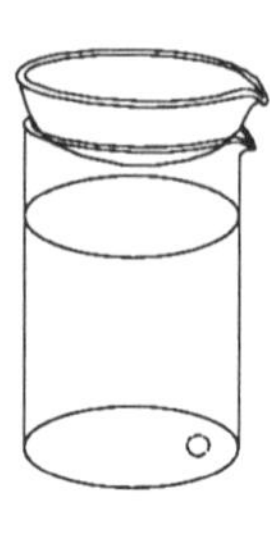

5. When you have obtained solutions of $CuSO_4 \cdot 5H_2O$ and NH_2CONH_2, the solvents must be evaporated as they were in Inquiry 4. However, in this Inquiry you have more solvent, so heat should be applied to speed up the process. The water solution can be evaporated on a steam bath (as illustrated) over a burner (see below in 6), or on a hot plate. However, the flammability of ethanol requires that the solution containing C_2H_5OH be evaporated without a flame and away from the burner used to evaporate the H_2O from the $CuSO_4 \cdot 5H_2O$. So if a steam bath is used to evaporate the ethanol, it should be made by heating a beaker of water to boiling, and then turning off the burner. An evaporating dish containing the ethanol-urea solution can be set on the beaker of hot water. The water will remain hot long enough to evaporate the ethanol. Note on your flowchart the type of vessel and the method of evaporation that will be used to obtain each solid.

6. Your instructor will demonstrate the following new techniques, and space is provided for you to take notes.

a. measurement of liquid volume using a graduated cylinder (note the *meniscus*):

b. decanting:

c. filtering:

d. rinsing a solid from one container to another using a wash bottle:

e. setting up a steam bath using an evaporating dish and beaker:

SAFETY NOTES

1. Wear approved safety goggles at all times in the lab.

2. Ethanol, C_2H_5OH, is FLAMMABLE. Do not use it near flames.

3. All chemicals must be handled carefully and treated with respect. The solids and liquids used in this Inquiry are safe for you to use responsibly.

4. Wash your hands before you leave the lab.

PROCEDURE

1. Tear out the Inquiry 5 Observations page, and have it ready.

2. Put a tiny sample of sand, SiO_2, in each of two small test tubes. To one add a few drops of distilled water, H_2O, and to the other a few drops of ethanol, C_2H_5OH. Is sand soluble in either of these two solvents? Record your observations, and be sure that you have put the sand in the appropriate place on your flowchart.

3. Have your instructor approve your flowchart before you begin. Turn in your procedure flowchart with your observations and results.

4. Watch your instructor make the mixture of copper(II) sulfate pentahydrate, $CuSO_4{\cdot}5H_2O$, urea, NH_2CONH_2, and sand, SiO_2.

5. Weigh out about 1 g of the mixture to the nearest 0.001 g and begin your separation.

6. When you finish, follow the instructions below for Cleanup.

CLEANUP

(All Cleanup suggestions are subject to local laws governing waste from laboratories. The following are suggestions only and may be changed by your instructor. Space has been provided for additional instructions.)

1. The materials you have used are not harmful to the environment in small quantities and, except for the sand, may be disposed of by rinsing them down the drain. Place the sand in the trash.

2. Return all chemicals to their proper places in the lab, and wash and dry your lab bench. Wash your hands before you leave.

3. Additional instructions:

Name ______________________________________Sec______________

INQUIRY 5: OBSERVATIONS

1. Write the number of your mixture:____________ (Fill this in only if your instructor makes up more than one mixture.)

2. Is sand soluble in water?___________ In ethanol?___________

3. Mass of mixture:

(Note: If your balance can be tared, your instructor will show you how to use the tare, and you can determine the mass of each solid directly, without determining the mass of the vessel.)

Mass of vessel + mixture	g
Mass of vessel	g
Mass of mixture	g

4. Mass of urea in mixture:

Mass of vessel + urea	g
Mass of vessel	g
Mass of urea, NH_2CONH_2	g

5. Mass of copper sulfate in mixture:

Mass of vessel + $CuSO_4 \cdot 5H_2O$	g
Mass of vessel	g
Mass of $CuSO_4 \cdot 5H_2O$	g

6. Mass of sand in mixture:

Mass of vessel + sand	g
Mass of vessel	g
Mass of sand, SiO_2	g

Name ______________________________ Sec__________

INQUIRY 5: RESULTS

1. Summarize your results from the Observations page:

Mass of urea, NH_2CONH_2:	__________g
Mass of copper(II) sulfate pentahydrate, $CuSO_4 \cdot 5H_2O$:	__________g
Mass of sand, SiO_2:	__________g
Total mass:	__________g

2. How does the total mass you found compare with the starting mass?

3. How do you account for any difference between the total mass and the starting mass?

Name ______________________________ Sec__________

INQUIRY 5: FOLLOW-UP QUESTIONS

These are to be done in the laboratory after the Inquiry. You are encouraged to discuss these with your lab partner and your lab instructor.

Your instructor has given you a white solid mixture in a test tube. It contains ammonium chloride, NH_4Cl, sodium chloride, NaCl, and sand, SiO_2. From a chemical handbook you have found the following:

a. NaCl is soluble in water. It melts at 801°C and boils at 1413°C.

b. NH_4Cl is soluble in water, and does not have a melting point. Instead, like dry ice (solid CO_2), it *sublimes*, i.e., when heated to 340°C, NH_4Cl becomes a gas without melting. It resolidifies on cooling.

c. SiO_2 is not soluble in water. It melts at about 1700°C.

d. The temperature reached in an open container on a Bunsen burner flame is approximately 750°C.

Using a flowchart, show *in detail* how you would determine the mass of each of the components in the mixture.

INQUIRY 6

Does a Reaction Occur?

Chemists look for changes in matter as an indication that a chemical reaction has occurred. For example, if two solutions are poured into a container, questions that might be asked are, "Does a precipitate form?" "Is a gas evolved?" "Is there a color change?" "Is heat evolved?" A positive answer to one of these questions is a good indication that a chemical reaction has occurred and that one or more new species are formed. In this Inquiry you will use your 96-well plate to make a matrix of compounds and examine their interactions.

Equipment you will need

microstirrer
test tubes (8), 75 or 100 mm
96-well plate

Chemicals you will use

Solutions, 0.1*M*, in dropper bottles

calcium nitrate, $Ca(NO_3)_2$
copper(II) nitrate, $Cu(NO_3)_2$
hydrochloric acid, HCl
iron(III) nitrate, $Fe(NO_3)_3$
potassium carbonate, K_2CO_3
potassium hydroxide, KOH
potassium iodate, KIO_3
silver nitrate, $AgNO_3$
sodium carbonate, Na_2CO_3
sodium chloride, NaCl
sodium hydroxide, NaOH
sodium phosphate, Na_3PO_4
sodium sulfate, Na_2SO_4
sulfuric acid, H_2SO_4

phenolphthalein solution (dropper bottle)

BEFORE YOU BEGIN

1. Most of the compounds we have worked with thus far are *ionic*, meaning that they contain a positive metal ion and a negative ion, each of which can be either a single atom or polyatomic. For example, you are familiar with sodium chloride, NaCl. Sodium is in Group I, and it loses one electron to form Na^+. Chlorine is in Group VII and gains one electron to form Cl^-. Since positive and negative charges must add up to zero, only one Na^+ ion and one Cl^- ion are necessary in the formula. Calcium in Group II loses 2 electrons and forms Ca^{2+}. The calcium ion,

therefore, requires 2 negative charges to form a neutral compound. It can be satisfied a number of ways, but it commonly forms compounds which have one ion with a -2 charge, such as sulfate, SO_4^{2-}, making $CaSO_4$, or two ions with -1 charge, such as chloride, Cl^-, to form $CaCl_2$.

2. In this Inquiry new compounds may be formed as two compounds are placed in a well of your 96-well plate. You must be able to determine *whether* something has occurred, and, if so, *what* has occurred, and you must write formulas and equations for the reactions. For example, assuming that the cations simply exchange negative ions (double displacement), complete the following in words and formulas:

sodium chloride + silver nitrate yields

______________________ + ______________________

$NaCl$ + $AgNO_3$ → ______________ + ______________

Sometimes a gas is produced, and from the carbonates in this Inquiry we might expect to get carbon dioxide, CO_2, when reaction with acid occurs. Given the following equation in words, complete and balance the equation with formulas:

sodium carbonate + hydrochloric acid yields sodium chloride + carbon dioxide + water

Na_2CO_3 + HCl → ______________________________

Carbonates can also form precipitates in double displacement reactions.

sodium carbonate + silver nitrate yields sodium nitrate + silver carbonate

Na_2CO_3 + $AgNO_3$ → ______________________________

3. Below is an array of compounds in a well plate. Assuming that these are double displacement reactions, write the correct formulas of the compounds that might be formed in each well. Two examples are given.

	1 $AgNO_3$	2 $Hg(NO_3)_2$
A NaCl	$NaNO_3$ + $AgCl$	
B K_2CO_3	KNO_3 + Ag_2CO_3	

SAFETY NOTES

1. Wear approved safety goggles at all times in the lab.

2. All chemicals must be handled carefully and treated with respect. The solids and liquids used in this Inquiry are safe for you to use responsibly.

3. Wash your hands before you leave the lab.

PROCEDURE

Tear out the Inquiry 6 Observations page and place it beside your well plate. Examine the page carefully, and you will see that the solutions which you are to use in each row and each column are already listed. Read the entire procedure before obtaining any reagents.

Part I: Using Test Tubes

Since silver ion tests must be carried out separately so that the waste can be easily collected, set up the "A" row of your matrix in 8 small test tubes. First put several drops of silver nitrate, $AgNO_3$, in each tube, and then add the solutions indicated in columns 1-8. On the Observations page note what occurs, and write formulas and names for the possible products. Dispose of the silver compounds as indicated in the Cleanup section.

Part II: Using the Well Plate

1. Now, using the 96 well plate, begin row "B" by putting a few drops of calcium nitrate, $Ca(NO_3)_2$, in each well in the row. Put copper(II) nitrate, $Cu(NO_3)_2$, in each well of the next row. Continue the rows until the last row has a few drops of H_2SO_4 in

each well. Complete the columns, and you will have two reactants in each well. Make your observations carefully. If a precipitate forms, write "ppt" and its color. If it bubbles, a gas is being produced. Remember to write names and formulas of possible products. If nothing happens in a well, write "NR" meaning "no reaction."

2. There are four wells in which the observations you make will likely be correct but misleading: D3, D7, F3, and F7. These are the wells in which the acids, HCl and H_2SO_4, are added to the bases, NaOH and KOH. When an acid reacts with a base, the result is a salt and water. Since the salt may be soluble in water and the amount of water produced, compared to the amount that was already in the well, is *very small*, there may be no visible indication that a reaction has occurred. To determine if a reaction has occurred we use an *indicator*, a dye which changes color when it reacts with an acid or a base. To check this, find a clean portion of your well plate and put a few drops of hydrochloric acid, HCl, in one well, and sulfuric acid, H_2SO_4, in another. Put a drop of the indicator, phenolphthalein, in each solution. Add sodium hydroxide, NaOH, dropwise to each one until a permanent color change takes place. When it occurs, the NaOH has neutralized the acid, forming a salt and water. Repeat with KOH. You may study this reaction quantitatively in a later Inquiry.

CLEANUP

(All Cleanup suggestions are subject to local laws governing waste from laboratories. The following are suggestions only and may be changed by your instructor. Space has been provided for additional instructions.)

1. The materials you have used, except silver compounds, are not harmful to the environment in small quantities and may be disposed of by rinsing them down the drain. Empty the test tubes containing the silver compounds into the waste container for recycling. Rinse the tubes with a few drops of water, and add the rinsings to the waste container.

2. Run plenty of water over your 96-well plate and shake it dry before returning it to the storeroom or putting it away.

3. Return all chemicals to their proper places in the lab, and wash and dry your lab bench. Wash your hands before you leave.

4. Additional instructions:

Name ______________________________ Sec________

INQUIRY 6: OBSERVATIONS

(cont'd next page)

	1 K_2CO_3	2 KIO_3	3 KOH	4 Na_2CO_3
A $AgNO_3$				
B $Ca(NO_3)_2$				
C $Cu(NO_3)_2$				
D HCl				
E $Fe(NO_3)_2$				
F H_2SO_4				

Name ______________________ Sec______

INQUIRY 6: OBSERVATIONS (cont'd)

	5 $NaCl$	6 Na_3PO_4	7 $NaOH$	8 Na_2SO_4
A $AgNO_3$				
B $Ca(NO_3)_2$				
C $Cu(NO_3)_2$				
D HCl				
E $Fe(NO_3)_2$				
F H_2SO_4				

Name ____ Sec______________

INQUIRY 6: RESULTS

In each well in which a change was observed, a chemical reaction occurred. Now, with a little thought, you can identify each product and write an equation expressing what occurred. For example, in test tube A1 you observed a white precipitate. You have already determined that the possible products of the reaction between NaCl and $AgNO_3$ were AgCl and $NaNO_3$. Is the white precipitate a mixture of the two products? Is it only one of them? If it is one of the products, which one is it?

To figure this out, go back to Inquiry 2 and look at the results of the solubilities of all the compounds containing the nitrate ion, NO_3^-. Are nitrates soluble or insoluble in water? ____________________ If you did not do Inquiry 2 there is enough information in this Inquiry to answer that question. Look at the list of solutions you used under "Chemicals you will use." If several of those solutions are nitrates, what does that imply about the solubility of nitrates in water?____________________

If $NaNO_3$ is soluble in water, the precipitate, the insoluble product, must have been AgCl. To indicate solubility in water we use *(aq)* following any soluble compound. (A compound that dissolves in water is said to be in *aqueous* solution. *Aqueous* comes from the Latin *aqua*, meaning "water.") Therefore, the equation for this reaction is written:

$$NaCl(aq) + AgNO_3(aq) \rightarrow NaNO_3(aq) + AgCl(s)$$

Silver chloride, AgCl, is a precipitate or insoluble *solid*, so we indicate that with *(s)*. As you might imagine, if a *gas* is a reactant or a product, we use *(g)* following the formula.

Now look at your Observations page. An array which has 6 rows and 8 columns has 48 wells. This could mean that you need to write 48 equations! But wait! In about 1/4 of the wells you should have written "NR" to indicate that nothing happened. If nothing happened, there is no equation. You will also find as you write equations that many are similar. For instance, sodium nitrate, $NaNO_3$, occurs several times. If potassium nitrate, KNO_3, is used or produced, use your knowledge of the periodic table and the fact that Na and K are in the same family to help you determine what KNO_3 should do.

Your instructor will help you balance equations and will show you how to handle the reactions that produce gases and those that had color changes without precipitates. On the next pages, write the balanced equation for the reaction that occurred in each well in which there was a change. If there was no reaction, write NR.

Name ______________________________ Sec____________

WELL	EQUATION
A1	
A2	
A3	
A4	
A5	
A6	
A7	
A8	
B1	
B2	
B3	
B4	
B5	
B6	
B7	
B8	
C1	
C2	
C3	

Name ______________________________ Sec ____________

WELL	EQUATION
C4	
C5	
C6	
C7	
C8	
D1	
D2	
D3	Write the equation for the neutralization reaction that the indicator showed you in Procedure II,2.
D4	
D5	
D6	
D7	Write the equation for the neutralization reaction that the indicator showed you in Procedure II,2.
D8	
E1	
E2	
E3	
E4	
E5	
E6	

Name ______________________________ Sec__________

WELL	EQUATION
E7	
E8	
F1	
F2	
F3	Write the equation for the neutralization reaction that the indicator showed you in Procedure II,2.
F4	
F5	
F6	
F7	Write the equation for the neutralization reaction that the indicator showed you in Procedure II,2.
F8	

Name ______________________________________ Sec______________

INQUIRY 6: FOLLOW-UP QUESTIONS

These are to be done in the laboratory after the Inquiry. You are encouraged to discuss these with your lab partner and your lab instructor.

1. Make four generalizations about the reactions you have observed.

   a.

   b.

   c.

   d.

2. A student places a few drops of lead nitrate solution, $Pb(NO_3)_2$, into a well and adds a few drops of sodium chloride solution, NaCl. A white precipitate results.

   a. Write the name and formula of the precipitate.

   b. Write the equation for the reaction, noting aqueous and solid species.

3. The following array is obtained in an Inquiry similiar to this one. Identify the *solid* or *gas* formed in each well by name and formula.

	1 $Ba(NO_3)_2$	2 HNO_3
A Na_2SO_4	white ppt forms ____________ ____________	NR
B K_2CO_3	white ppt forms ____________ ____________	a gas is evolved ____________ ____________

Name ______________________________________ Sec______________

INQUIRY 6: FOLLOW-UP QUESTIONS (cont'd)

4. Write two questions inspired by this inquiry.

Unit 2

The Copper Story

Copper: A Representative Metal

There is nothing about copper, Cu, that is particularly unusual, unless perhaps it's the beautiful bright orange-gold color, or maybe the light blue of the copper(II) ion or the deep, electric blue of the copper-ammonia complex ion. It may be that it is distinctive in being one of the best conductors of electricity of all the metals; in fact, only silver is a better conductor than copper.

Of the elements, copper is 25th in abundance in the Earth's crust and is found in small concentrations in seawater. Its properties of malleability and ductility and the ease of extracting it from its ore have made copper a popular metal for vessels and jewelry for 5000 years. Turquoise, a hydrated form of copper aluminum phosphate, was mined thousands of years ago in the Sinai desert, where, atop a mountain called Serabit el-Khadem, there is an inscription that dates from the XIIth Egyptian dynasty (approximately 1800 BCE,* the Bronze Age). An Egyptian miner inscribed in stone his difficulty in finding the perfect turquoise in the hot summer: "The mountains are [as if] heated with red-hot iron and the colors [of the stones] are spoiled by it."† In one of the Inquiries to follow you will have an opportunity to form a hypothesis based on observations you will make, a hypothesis which could have helped the ancient miner understand why the colors of the turquoise might be altered by the hot sun.

You will find copper in the periodic table in the *transition metals*. Copper is element number 29, its atomic weight is 63.5 amu, and its density of 8.92 g/cm^3 is about average for the metals. The transition metals stand out for several reasons: (1) they are, for the most part, the structural metals of our world; (2) they form colored compounds of

---

* BCE means "Before the Common Era." This designation is slowly replacing the more parochial BC. AD then becomes CE, the Common Era.

† Kirsopp Lake, A. Barrois, Silva New, Romain F. Butin. "The Serabit Expedition of 1930," *Harvard Theological Review*, XXV (1932), 95-203. (Barrois confirms "the instability of the Sinai turquoise which easily passes from celestial blue to green, and from green to a dull grey." (p. 114).

unequaled beauty; (3) three of them, copper, silver, and gold, have been the coinage metals since antiquity.

Copper in the Laboratory

Other metals are sometimes used in the Inquiries that follow, but copper is chosen as the principal metal to study, not for its uniqueness, but because it is *representative* of the metals. All metals undergo reactions that are unique to the individual metal, but metals as a class of elements have many types of reactions in common. We will use copper and its compounds to help us understand some very basic areas of chemistry, such as formulas of compounds, reactions, acids and bases, and oxidation-reduction. And we'll take a closer look at the familiar penny, examining it chemically this time.

We choose to use copper for another reason as well. Copper is one of the elements which has fairly low toxicity to humans. It is essential in our diets (we normally consume a few milligrams daily), but in large quantities it can be toxic. As usual, you must handle its compounds responsibly and dispose of them as you are directed. Since copper is not very toxic in small quantities, small amounts of copper ion, Cu^{2+}, can go down the drain where local laws allow; where necessary, it can be recycled in one of its compounds.

INQUIRY 7

What Is a Metal?

Although we are studying copper as a representative metal, let us look at metals in general to see what physical properties they have in common and what types of reactions they undergo.

Equipment you will need

beakers, several small
conductivity tester or multimeter
droppers
graduated cylinder, 10 mL
stirring rod
test tubes, small
tongs
24- or 96-well plate

Other items

litmus paper, red and blue
paper towel
steel wool for polishing metal pieces
various metallic and nonmetallic elements for classification

Chemicals you will use

Solids
- calcium oxide, CaO
- copper(II) oxide, CuO
- iron(III) oxide, Fe_2O_3
- metal pieces, 5 x 5 mm
  - aluminum, Al, calcium, Ca, copper, Cu, iron, Fe (or steel wool), magnesium, Mg (2 cm), zinc, Zn
- sodium carbonate, Na_2CO_3
- zinc oxide, ZnO

Solutions
- hydrochloric acid, HCl, 3*M*
- nitric acid, HNO_3, conc.
- phenolphthalein solution
- sodium hydroxide, $NaOH$, 3*M*

BEFORE YOU BEGIN

1. In Inquiry 6, Procedure II,2, you were introduced to a neutralization reaction, where an acid neutralizes a base, or vice versa. The result was a salt and water. Write the equation for the reaction of hydrochloric acid with sodium hydroxide.

2. Your instructor will demonstrate how to test element samples in Procedure I for *malleability*, *ductility*, and *conductivity*. Define these terms.

SAFETY NOTES

1. Wear approved safety goggles at all times in the lab.

2. Hydrochloric acid, HCl, nitric acid, HNO_3, and sodium hydroxide, NaOH, are CAUSTIC! Use HNO_3 only in the hood; do not take the bottle to your bench. If you get any of these chemicals on your skin or clothing, wash immediately with running water and tell your instructor.

3. All chemicals must be handled carefully and treated with respect. The solids and liquids used in this Inquiry are safe for you to use responsibly.

4. Wash your hands before you leave the lab.

PROCEDURE

Part I: Metal or Nonmetal?

1. Tear out the Inquiry 7 Observations page, and have it ready.

2. Your instructor will have some elements out on the bench for you to examine and classify. Follow the additional instructions from your instructor, and test the elements for malleability and conductivity. Look for evidence of ductility. Take careful observations.

Part II: Some Reactions of Metals

A. Reactions with acid

1. Obtain about 5 mL of hydrochloric acid, HCl, in a small beaker and, on a piece of paper, one each of the small metal pieces listed in "Chemicals you will use." Using a dropper, place a dropperful of HCl in each of 6 wells in the 24-well plate, or in individual small test tubes. Before you add the metal pieces, be prepared to observe *whether* a reaction occurs, note the *rate* of the reaction (i.e., how fast the reaction occurs) and rank the metals by fastest to slowest to react.

2. Polish the zinc, aluminum, and iron pieces with steel wool. Add a small piece of metal to the acid in each well. Record your observations on the Inquiry 7 Observations page.

3. This step must be done IN A HOOD. Stand a small test tube in a small beaker, and take this apparatus to the hood. Put a tiny piece of copper in the test tube. Add a drop of concentrated nitric acid, HNO_3 (CAREFUL!), to the copper. What happens? DO NOT BREATHE THE GAS THAT IS PRODUCED. Leave the beaker in the hood until you have finished the rest of the Inquiry. Then pour the remaining solution into the waste beaker as indicated in Cleanup. Rinse the test tube.

B. Metal oxides

1. Cover the bottom of a small beaker with a few milliliters of water. Place it next to a burner, so that the product of the following reaction can fall into the water. Obtain a piece of magnesium, Mg, about 2 cm long. Grasping the Mg with tongs or forceps, hold it in a burner flame until it ignites, and then hold it over the beaker. DO NOT LOOK DIRECTLY AT THE BURNING MAGNESIUM! Drop the product into the water.

2. Dissolve the solid in the water, breaking it up with a stirring rod. Add a drop of phenolphthalein to the mixture. Record your observations. Make a reference standard by putting a drop of hydrochloric acid, HCl, in a well of the well plate, and a drop of sodium hydroxide, NaOH, in another well. Add a drop of phenolphthalein to each well. What do you see? HCl is an *acid*, and NaOH is a *base*. What type of compound is the oxide of magnesium that you made? What is its formula?

3. Using the 24- or 96-well plate, place a very small amount of iron(III) oxide in each of two wells. Do the same thing with copper(II) oxide, calcium oxide, and zinc oxide. Add a dropperful of HCl to each oxide. (If the result of adding acid to the Fe_2O_3 and the CuO is not obvious, withdraw a little of the solution into a dropper and look at its color. What does the color tell you? Can you tell if the oxide has dissolved in the acid?) Record your observations.

CLEANUP

(All Cleanup suggestions are subject to local laws governing waste from laboratories. The following are suggestions only and may be changed by your instructor. Space has been provided for additional instructions.)

1. Pour all liquids from reaction vessels into a large beaker. If any solids remain in the reaction vessels, they may be put in the trash after they are rinsed with water. Check the liquid waste with litmus paper to determine if it is acid. If so, neutralize it as follows. Place the beaker of waste acid in the sink and slowly add sodium carbonate, Na_2CO_3, or sodium bicarbonate, $NaHCO_3$, solution. When the addition of carbonate or bicarbonate solution no longer causes "fizzing," flush the resulting solution down the drain with running water. If the liquid waste is basic it may be poured down the drain with running water.

2. Return all chemicals to their proper places in the lab, and wash and dry your lab bench. Wash your hands before you leave.

3. Additional instructions:

Name ______________________________ Sec__________

INQUIRY 7: OBSERVATIONS

Part I

Record your observations of the demonstration.

Part II

A. Reactions with acid

1. For each metal used, write the name of the metal, and what you observed.

   What is the gas that is being produced?

2. Put the metals in order of the *rate of reaction* with HCl, fastest to slowest. The way we usually show this is to use the *greater than* symbol, >, from mathematics. So if we want to show that a car is faster than a bicycle, and a bicycle is faster than a tricycle, we would write :

   rate of travel: car > bicycle > tricycle

3. What happened when you put copper, Cu, in HNO_3?

   What is the gas which is produced? If you don't know, look up nitric acid or copper in your text.

Name ______________________________ Sec____________

INQUIRY 7: OBSERVATIONS (cont'd)

B. Metal oxides

Record your observations for steps 1, 2, and 3.

Name ______________________________________ Sec______________

INQUIRY 7: FOLLOW-UP QUESTIONS

These are to be done in the laboratory after the Inquiry. You are encouraged to discuss these with your lab partner and your lab instructor.

1. If you were given an unknown solid and asked to determine if it were a metal or a nonmetal, what steps would you take? Would a single test, such as electrical conductivity, be sufficient?

2. What generalization can you make about the oxides of metals?

3. Write and balance equations for each of the following.

   a. magnesium reacts with oxygen to form magnesium oxide

   b. magnesium oxide + water produces magnesium hydroxide

   c. iron + oxygen $\rightarrow$ iron(III) oxide

   d. chromium(III) oxide + hydrochloric acid produces chromium(III) chloride + water

   e. zinc + hydrochloric acid $\rightarrow$

   f. copper + nitric acid yields copper(II) nitrate + nitrogen dioxide + water

Name ______________________________ Sec__________

INQUIRY 7 : FOLLOW-UP QUESTIONS (cont'd)

4. a. Based on your observations, if you had to store hydrochloric acid in a metal container, which metal would you choose for the container?

   b. Why?

5. Write two questions that are inspired by this Inquiry.

   a.

   b.

INQUIRY 8

A Circle of Copper Compounds

Starting with copper(II) sulfate pentahydrate, $CuSO_4 \cdot 5H_2O$, a series of reactions illustrate the ease with which some metal compounds can be formed and destroyed, finally reproducing the original material. The product of this Inquiry is saved to become the starting material for the next Inquiry.

Equipment you will need

beakers, 100 and 600 mL
Bunsen burner or hot plate
crucible
funnel with filter paper
ring stand with ring
stirring rod
vial
wash bottle
wire gauze
wire triangle

Chemicals you will use

copper(II) sulfate pentahydrate, $CuSO_4 \cdot 5H_2O$
ethanol, C_2H_5OH
hydrochloric acid, HCl, 6*M*
sodium carbonate, Na_2CO_3, 1*M*
sodium hydroxide, NaOH, 6*M*
sulfuric acid, H_2SO_4, 3*M*

Other items

litmus paper, red and blue

BEFORE YOU BEGIN

1. The salt used in this Inquiry, $CuSO_4 \cdot 5H_2O$, is called a *hydrate*. Look up hydrates in the index of your textbook and give two other examples below.

   a.

   b.

2. What is the meaning of the "dot" before the water molecules?

3. What is the *common name* of copper(II) sulfate pentahydrate, $CuSO_4 \cdot 5H_2O$?

4. Your instructor will demonstrate the following.

a. the proper way to set up and heat a crucible:

b. how to set up a conical funnel with filter paper:

SAFETY NOTES

1. Wear approved safety goggles at all times in the lab.

2. The following acids and bases are CAUSTIC and must be used with care:

hydrochloric acid, HCl, 6*M*
sodium hydroxide, NaOH, 6*M*
sulfuric acid, H_2SO_4, 3*M*

These can cause severe burns. If you get any of these on your skin or clothing, rinse with running water *immediately* and tell your instructor.

3. All chemicals must be handled carefully and treated with respect. The solids and liquids used in this Inquiry are safe for you to use responsibly.

4. Copper compounds are toxic if ingested in the amounts used in this Inquiry. Handle them carefully.

5. Wash your hands before you leave the lab.

PROCEDURE

Part I: Initial Observations of the Starting Material

1. Tear out the Inquiry 8 Observations page, and set it up as indicated.

2. Put a small amount of copper(II) sulfate pentahydrate, $CuSO_4 \cdot 5H_2O$, in a crucible and examine it closely. Weigh the crucible + salt. Write your observations on the Observations page.

3. Place the open crucible containing the salt on a ring stand as demonstrated. Begin heating the crucible gently, and then heat more intensely. What happens?

4. When there is no more change, stop heating and let the crucible cool. While it is cooling, begin Part II. After the crucible is cool, weigh it, and then add a few drops of water and observe.

Part II: The Circle of Compounds

1. Weigh a 100-mL beaker, record its mass, and then add about 1 g of copper(II) sulfate pentahydrate, $CuSO_4 \cdot 5H_2O$. Weigh the beaker + salt to the nearest 0.001 g, recording the mass on the Observations page.

2. Add about 5 mL of water, and stir to dissolve the salt. What color is the solution? If the salt is slow to dissolve, place the beaker on a wire gauze on a ring stand, and heat the solution gently with a Bunsen burner.

3. Allow the solution to cool, and then add 5 mL of 1*M* sodium carbonate, Na_2CO_3. Record what happens. What is the product?

4. Obtain about 3 mL of 6*M* hydrochloric acid, HCl, and, using a dropper, add a few drops of the acid to the precipitate. What happens? What do you think is the gas that is being produced? Slowly and carefully add HCl with stirring until the precipitate is dissolved.

5. Obtain about 10 mL of 6*M* sodium hydroxide, NaOH. Add it with stirring to the beaker of solution. Describe the product.

6. Heat the mixture gently with stirring. Do not stop stirring, because bubbles tend to form under the precipitate, and they can rise violently to the surface, bumping solution out of the beaker. Stirring keeps large bubbles from forming. What is happening to the precipitate?

7. When the color change is complete, stop heating and stirring, and allow the precipitate to settle. This may take 5 or 10 minutes. What color is the supernatant liquid? Carefully decant this solution into a 600-mL beaker, retaining the precipitate in the 100-mL beaker. The solution in the large beaker is waste; set it aside to collect other waste.

8. Slowly add 3-5 mL of 3*M* sulfuric acid, H_2SO_4, with constant stirring. Use the smallest amount of acid possible, but if solution is not complete, add acid until it is complete. What color is the solution now? What is the product?

9. Evaporate the liquid over a gentle flame until the solid begins to crust around the edges. Remove the flame and cool the solution by setting the beaker in a larger beaker of cold water.

10. Add 10 mL of ethanol, C_2H_5OH, and stir. What is happening? What have you observed in previous Inquiries about the solubility in ethanol of copper(II) sulfate pentahydrate, $CuSO_4 \cdot 5H_2O$?

11. After the crystals have formed, pour off the supernatant liquid into the waste beaker. Rinse the crystals with a little more C_2H_5OH. Pour the rinsings into the waste beaker. See Cleanup below for waste instructions.

12. Set up a funnel with a weighed filter paper. Now add a few additional milliliters of C_2H_5OH to the crystals, stir the crystals and dump them onto the filter paper. If some crystals remain in the flask, add a few milliliters more of C_2H_5OH, again stir, and

quickly pour the crystals onto the filter. Pour the filtrate into the waste container beaker.

13. Open the filter paper and let the ethanol evaporate. When crystals and paper are dry, weigh the product. Place the dry $CuSO_4 \cdot 5H_2O$ in a vial and stopper it. Label it, and save it for the next Inquiry.

CLEANUP

(All Cleanup suggestions are subject to local laws governing waste from laboratories. The following are suggestions only and may be changed by your instructor. Space has been provided for additional instructions.)

1. Rinse the crucible with water and dry.

2. Check the liquid in the waste beaker to see if it is nearly neutral. If so, or if it is basic, pour it down the drain with water. If it is acid, neutralize the excess acid as follows. Place the beaker of waste acid in the sink and slowly add sodium carbonate, Na_2CO_3, or sodium bicarbonate, $NaHCO_3$, solution. When the addition of carbonate or bicarbonate solution no longer causes "fizzing," flush the resulting solution down the drain with running water.

3. Save your $CuSO_4 \cdot 5H_2O$ crystals for the next Inquiry in a tightly stoppered vial.

4. Return all chemicals to their proper places in the lab, and wash and dry your lab bench. Wash your hands before you leave.

5. Additional instructions:

Name ______________________________ Sec ____________

INQUIRY 8: OBSERVATIONS

Almost every step of the Procedure requires that you make observations. Set up your Observations page to record weighings as you have done in earlier Inquiries, and answer the questions that are in the Procedure. Number your observations to correspond to steps in the Procedure. Write the names and formulas of products formed in every case.

Name ______________________________ Sec__________

INQUIRY 8: RESULTS

1. Explain what occurred when $CuSO_4 \cdot 5H_2O$ was heated in the crucible in Part I.

2. Write equations for each step in Part II in which a compound was formed.

Name ________________________________ Sec____________

INQUIRY 8: FOLLOW-UP QUESTIONS

These are to be done in the laboratory after the Inquiry. You are encouraged to discuss these with your lab partner and your lab instructor.

1. If you started with 1.000 g of $CuSO_4 \cdot 5H_2O$, how many grams of dry CuO could you expect to make? (Hint: Look at the equations you have written on p. 66, and determine the mole relationship between the starting material and CuO.)

2. Brad begins an Inquiry by reacting 10 g of iron (III) chloride hexahydrate with an excess of potassium hydroxide. His partner, Dawn, then adds excess sulfuric acid, H_2SO_4, to dissolve the iron(III) hydroxide. Brad then adds sodium iodate, $NaIO_3$, to make iron(III) iodate.

   a. Write an equation for each step of Brad's and Dawn's experiment.

   b. If Dawn dries and weighs the iron(III) iodate, how much should she get?

Name ______________________________ Sec__________

INQUIRY 8: FOLLOW-UP QUESTIONS (cont'd)

3. Turquoise, mentioned in the introduction to Unit 2, p. 51, can have the formula, $CuAl_6(PO_4)_4(OH)_8 \cdot H_2O$. Based on your observations of $CuSO_4 \cdot 5H_2O$ in Part I and your explanation for those observations, write a hypothesis to explain why the ancient Egyptian miner mentioned in the introduction to Unit 2 might have found discolored turquoise in the desert summer.

INQUIRY 9

Redox: Transferring Electrons

In previous Inquiries we have examined different types of reactions, including acid-base and precipitation reactions. In Inquiry 7, we saw that some metals are consumed by hydrochloric acid, producing a gas in the process, and that one metal, magnesium, Mg, burns brightly in oxygen, O_2, forming a new, nonmetallic substance, an oxide. Reactions in which metals react to form products that are no longer metallic in nature belong to a large class of reactions in which electrons are transferred from one element to another. These are called *oxidation-reduction* or "redox" reactions. Although redox reactions can occur between nonmetals as well as between metals and nonmetals, we will confine this Inquiry to metals and metal-ion solutions, because these are easy to handle. The principles learned here can be applied to many chemical systems.

Equipment you will need

battery tester or voltmeter
beakers (5), 50 mL
beaker, 600 mL
dissecting microscope (optional)
oven (optional)
test tubes (several), 100 mm
watch glass
24-well plate
wire leads with alligator clips

Other items

filter paper
magnifying glass
steel wool

Chemicals you will use

Solids
- calcium turnings, 1 piece
- copper(II) sulfate pentahydrate, $CuSO_4 \cdot 5H_2O$ (from Inquiry 8)
- copper wire, Cu, 16 gauge, 25 cm
- iron nails, Fe (or brads)
- magnesium ribbon, Mg, 1 cm
- metal strips, 5 x 0.5 cm (Cu, Zn, Fe)
- Zn metal pieces, 5 x 5 mm

Solutions
- ammonium chloride, NH_4Cl, sat'd
- hydrochloric acid, HCl, 0.1*M*
- iron (III) nitrate, $Fe(NO_3)_3$, 0.1*M*
- phenolphthalein solution
- silver nitrate, $AgNO_3$, 0.1*M*
- sodium chloride, NaCl, 0.1*M*
- sodium hydroxide, NaOH, 0.1*M*
- zinc nitrate, $Zn(NO_3)_2$, 0.1*M*

water, distilled, H_2O

BEFORE YOU BEGIN

1. The Silver Snake and the corrosion of iron below require setting up at least a day before the Inquiry is done.

2. Look up "dendrite" in the dictionary. If you can't find it, look for the root, "dendr—."

3. Your instructor will demonstrate the U-tube you will use to make a salt bridge.

SAFETY NOTES

1. Wear approved safety goggles at all times in the lab.

2. Silver nitrate causes a stain on skin and clothing. If you get this solution on your hands or clothes, rinse well with water.

3. All chemicals must be handled carefully and treated with respect. The solids and liquids used in this Inquiry are safe for you to use responsibly.

4. Wash your hands before you leave the lab.

PROCEDURE

Part I: The Day or Week Before

A. The Silver Snake

1. Make a spiral of a piece of Cu wire so that the spiral will occupy about half of a 50-mL beaker.

2. Weigh the Cu wire and the 50-mL beaker separately, and record their masses.

3. Place the spiral of wire in the beaker, and cover the wire with a solution of 0.1*M* silver nitrate, $AgNO_3$. Cover the beaker with a watch glass, and put it away in your lab locker until the next lab period.

B. The corrosion of iron

1. Place one shiny iron nail or brad into each of four 100-mm test tubes. Into one put a few milliliters of distilled water; into the next, hydrochloric acid solution; into the third, sodium hydroxide solution; and in the fourth, sodium chloride solution.

2. In a fifth tube place sodium chloride solution and a nail wrapped with a zinc wire or strip.

3. Let these five test tubes sit overnight or over the week, and examine them when you come to lab. What observations can you make about the effects of certain substances on iron corrosion? Dispose of the solutions in the waste beaker described in Part II.

Part II: The Silver Snake and Other Reactions

For the rest of the Inquiry, set up a 600-mL beaker for liquid waste. Pour all liquid waste in this beaker. See Cleanup for disposal.

A. Examining the spiral

1. Next lab period, carefully lift the beaker and examine the result with a magnifying glass. If a dissecting microscope is available, gently move your beaker to the microscope and look at the crystals. What are the crystals which have formed? Describe them on your Observations page.

2. Remove any remaining Cu wire, rinsing material clinging to the wire back into the liquid. What color is the solution now? Dry and weigh the Cu wire.

3. Decant the solution into a clean 50-mL beaker. Set the solution aside for part II,B,1. Rinse the remaining solid several times with tiny portions of water, decanting the rinsings into the liquid waste beaker. Set the crystals aside to dry, either in an oven or in your locker, according to your instructor. When they are dry, weigh the crystals in the beaker. Dispose of them in the silver recycling container.

B. More reactions

1. Pour some solution from Part II,A,3 into a 2 wells of your 24-well plate, or into two small test tubes. Add a small piece of zinc to one and a piece of iron to another. Make initial observations, and then leave these to observe at the end of the lab period.

2. To check whether redox reactions are reversible, pour a few milliliters of zinc nitrate solution, $Zn(NO_3)_2$, into a clean test tube, and a few milliliters of iron(III) nitrate, $Fe(NO_3)_3$, into another. Put a small strip of copper, Cu, in each tube. Make initial observations, and leave these also for observation at the end of the lab period. After you have observed them, pour the waste solutions into the waste beaker.

Part III: Some Reactions with Water

1. Put a few milliliters of distilled water in two clean wells of the 24-well plate. In one, place a small piece of calcium metal, Ca, and in the other, a small piece of magnesium, Mg. Put a drop of phenolphthalein in each well. What happens? Continue to observe off and on during the period.

2. Try the same reactions in water with Cu and Zn.

Part IV: Taking Advantage of Electron Transfer

1. Make a solution of copper(II) sulfate pentahydrate, $CuSO_4 \cdot 5H_2O$, using either your product from Inquiry 8 or about 0.5 g of solid in about 20 mL of distilled water in a 50-mL beaker. Stir until it is dissolved. Warm if necessary to speed solution.

2. Put about 20 mL of 0.1*M* zinc nitrate, $Zn(NO_3)_2$, in another 50-mL beaker, and put the beakers side by side.

3. Place a strip of copper in the copper sulfate solution and a strip of zinc in the zinc nitrate solution. Obtain two wire leads with alligator clips on each end. Attach the end of one to the copper strip and its other end to the voltmeter. Do the same with the other wire to attach the zinc strip to the bulb or meter. What happens? Leave the apparatus wired as it is, and go on to either step 4 or 5 (depending on the type of salt bridge your instructor said to make.)

4. Obtain a small U-shaped glass tubing to use as a salt bridge. Fill it with ammonium chloride, NH_4Cl, from a wash bottle, and add a small plug of cotton which has been soaked in the NH_4Cl solution to each arm of the tube. Invert the salt bridge in the solutions. What happens? What have you made? Check your textbook for its proper name.

5. Obtain a piece of filter paper to be used as a salt bridge. Fold it into a 1/2-inch strip, and soak it in the ammonium chloride, NH_4Cl. Make a "U" of the filter paper into both 50-mL beakers, so that it forms a bridge between them. What happens? What have you made? Check your textbook for its proper name. Place all solutions in the waste beaker for Cleanup.

CLEANUP

(All Cleanup suggestions are subject to local laws governing waste from laboratories. The following are suggestions only and may be changed by your instructor. Space has been provided for additional instructions.)

1. Deposit the silver waste in the waste container provided so that it can be recycled.

2. All the metal pieces must be either recycled to waste jars, so that they can be cleaned to be reused, or put in the trash. Your instructor will tell you which to do.

3. Check the liquid waste to see if it is acid. If so, put the beaker in the sink and add Na_2CO_3 until the fizzing stops. Pour the resultant solution down the drain with running water.

4. Return all chemicals to their proper places in the lab, and wash and dry your lab bench. Wash your hands before you leave.

5. Additional instructions:

Name ________________________________ Sec____________

INQUIRY 9: OBSERVATIONS

Set up your Observations page to take observations and answer questions from each part of the Inquiry.

Name ______________________________ Sec__________

INQUIRY 9: RESULTS

1. When zinc metal loses 2 electrons to form Zn^{2+} ion, the *half*-reaction is written

   $Zn \rightarrow Zn^{2+} + 2e^-$ This is called ______________.

   When copper ion gains 2 electrons to form copper metal, the half-reaction is

   $Cu^{2+} + 2e^- \rightarrow Cu$. This is called ______________.

   The net oxidation-reduction reaction is $Zn + Cu^{2+} \rightarrow Zn^{2+} + Cu$. Note that charge is conserved, as is mass.

   For each of the reactions in Part II write a similar net equation on another sheet of paper, showing the half-reactions as above.

2. For the Silver Snake, calculate the moles of silver produced and the moles of copper used. Write the equation for this reaction, as above, and explain how the Silver Snake confirms that charge and mass are conserved.

3. Why do you suppose that the silver crystals you produced are said to be in the form of "silver dendrites"?

4. What conclusions can you draw about corrosion of iron? What was the purpose of the zinc strip on one of the nails? Look up "galvanizing" in your text and write the definition.

5. a. What gas was released when calcium reacted with water?

   b. What did the phenolphthalein indicate was also formed?

   c. Write an equation to represent the reaction of calcium with water.

   d. Write an equation for the reaction of Mg with water.

Name ______________________________ Sec__________

INQUIRY 9: RESULTS (cont'd)

6. Draw a diagram of the cell you made in Part IV. Label all parts and show the direction of electron movement. Your text should help.

7. Look at the observations which involve Zn, Cu, and Fe. Which element is the least easily oxidized? Or, to put it another way, which element is most easily reduced?

   Can you decide which of the remaining two elements is more easily oxidized?

   Design a procedure which will answer that question.

Name ______________________________________ Sec______________

INQUIRY 9: FOLLOW-UP QUESTIONS

These are to be done in the laboratory after the Inquiry. You are encouraged to discuss these with your lab partner and your lab instructor.

Although silver is much too expensive to use in a cell such as that in Part IV, a perfectly good cell could be made which utilizes the two half-reactions that make up the Silver Snake. Decide what will be oxidized and what will be reduced, and draw a cell using the materials of the Silver Snake and any other solution(s) and materials you wish to use to use. Show electron flow.

INQUIRY 10

Sunscreen: Good Cents!

As you discovered earlier, the modern penny is 0.5 g lighter than the penny it replaced in 1982 when a change was made in the materials of which it is composed (i.e., most of the copper in the penny was replaced by a zinc core). Based on the difference in reactivity of copper and zinc with hydrochloric acid (Inquiry 7), it is possible to remove the thin outer sheet of copper in which the zinc underlayer is clad. Zinc then becomes the starting material for the production of an everyday item, zinc oxide, used in some sunscreens, and the copper cladding becomes a conversation piece!

Equipment you will need

balance
beakers, various
Büchner funnel
Bunsen burner
crucible and lid
file
forceps, plastic
mortar and pestle
oven (optional)
ring stand and ring
spatula
suction flask and vacuum hose
wash bottle
watch glass
weighing dish or paper
wire hooks (for developing jar)
wire triangle

Chemicals you will use

ammonia, $NH_3(aq)$, 15*M*
hydrochloric acid, HCl, 6*M*
sodium carbonate, Na_2CO_3, 1*M*

Other items

commercial sunscreens (several, including one containing ZnO, if possible)
diazo paper (blueprint paper)
filter paper
large jar with lid
paper clip
penny, 1983 or later*
petroleum jelly
transparency film or plastic wrap
wax marking pencil

* It is not illegal to destroy a penny. The only caveat is that the product of the reaction, sunscreen, is not legal tender!

BEFORE YOU BEGIN

Your instructor will demonstrate some techniques needed for this Inquiry. Take notes in the spaces provided.

a. setting up a Büchner funnel:

b. transferring a precipitate to the funnel and rinsing it:

c. heating, cooling, and weighing a crucible to constant mass:

SAFETY NOTES

1. Wear approved safety goggles at all times in the lab.

2. Hydrochloric acid, HCl, is CAUSTIC and must be handled with extreme care. The 6*M* acid used in this Inquiry can cause severe burns. If you should spill any acid on your skin or clothes, wash with running water *immediately*, and inform your instructor. The ammonia vapor from concentrated aqueous ammonia, $NH_3(aq)$, is VERY IRRITATING; $NH_3(aq)$, must be used in the hood.

3. All chemicals must be handled carefully and treated with respect. The other solids and liquids used in this Inquiry are safe for you to use responsibly.

4. The sunscreen you make must not be put on your skin. Use it only as you are directed in the Procedure, and dispose of it in the trash at the end of the Inquiry.

5. Wash your hands before you leave the lab.

PROCEDURE

Part I: Removing the Copper (To be done at least 1 day before Part II.)

1. Tear out the Inquiry 10 Observations page, and have it ready.

2. Obtain a penny, 1983 or later.

3. Using a file, scrape a little copper off the edge of the penny around its circumference to expose the zinc layer beneath. Rinse and dry the penny. Weigh the penny and record its mass.

4. Place the penny in 25 mL of 6*M* hydrochloric acid, HCl, in a 100-mL beaker, and leave it overnight.

Part II: Zn to $ZnCO_3$ to ZnO (First Week)

1. Examine the penny. What has happened?

2. With plastic forceps, pick up the copper cladding and, holding it over the solution, rinse it with a stream of water so that the rinsings run back into the beaker. Set it aside, and weigh it when it is completely dry.

3. To the solution containing the zinc ion, add 50 mL of 1.0*M* Na_2CO_3. Stir well, adding some water, if necessary. What is formed?

4. Weigh a filter paper for use in the Büchner funnel, and record its mass.

5. Set up a Büchner funnel as demonstrated by your instructor, setting the filter paper in place with a little water. Obtain 100 mL of distilled water in a beaker, and have it ready for rinsing the precipitate. Turn the water on the aspirator down so that it is barely flowing, and transfer the precipitate to the filter paper in the funnel. It is important to filter the precipitate slowly so that it does not form cracks that would make rinsing inefficient, so try to keep the precipitate from running dry. As the rinse water reaches a level about 1/8 inch above the solid, add about 25 mL of water. Repeat 3 times, and after the 4 rinses, let the precipitate dry for several minutes at full vacuum. What ions are being rinsed from the precipitate?

6. Break the vacuum as your instructor demonstrated, and remove the Büchner funnel from the flask. Using a spatula, carefully transfer the filter paper and precipitate to a weighed watch glass. Write your initials on a small scrap of paper, and slip the paper under the filter paper so that you can identify your sample. Place the watch glass in a warm oven, or let it air dry overnight. When the solid is dry, remove the slip of paper with your initials, weigh solid, filter paper, and watch glass, and record the mass.

7. Carefully transfer the dry solid from the filter paper to a mortar, and, using a pestle, grind the solid to a powder.

8. Weigh a clean crucible and top, transfer the solid to the crucible, and reweigh.

9. Set a Bunsen burner under a wire triangle on a ring stand. Put the crucible on the wire triangle, and place the crucible top at an angle, half off the crucible, resting on the triangle. Heat the crucible gently at first, and then more intensely. Finally heat strongly for 5 minutes, so that the bottom of the crucible glows red. Remove the heat, set the top on the crucible, cool, and weigh.

10. Reheat for 5 minutes, cool, and reweigh. Continue the cycle of heating, cooling, and weighing, until constant mass is obtained. What is happening to the volume of the solid in the crucible? What is the product?

Part III: Sunscreen (Second Week)

Your instructor will set up the developing chamber by placing 10 mL of 15 M $NH_3(aq)$ in a large jar in the hood. The jar should have a loosely fitting top, and wire hooks should be hanging down in the jar from the rim.

1. Place about 1/3 of the solid ZnO you have made in a clean mortar and grind it to a fine powder.

2. Scoop a portion of petroleum jelly (about half a teaspoonful) onto a weighing dish or a piece of weighing paper. Add the finely ground solid ZnO to the jelly, and mix with a spatula. Describe the result.

3. Obtain a piece of precut transparency film or plastic wrap, and use a wax marking pencil to divide it into three equal areas. In one area, spread a small portion of commercial sunscreen. In the second area, spread an equal portion of your sunscreen. Leave the third area blank as a control.

4. With everything ready so that the following can be done quickly, obtain a piece of light-sensitive (diazo) paper, and immediately set the transparency film on it. Put the paper in the sun or on a windowsill (it does not require direct sunlight).

5. The paper will lose its yellow color, becoming almost white. When the paper is fully exposed (about 2 minutes), remove the plastic layer, and quickly hang the light-sensitive paper on a wire hook in the developing chamber in the hood. The developing takes only a few seconds. When the image on the paper turns dark blue, the developing is finished. Remove the paper from the chamber and lay it in the hood for a few minutes to let the ammonia odor dissipate. Wrap your diazo "photo" in a fresh piece of plastic wrap, and tape it on your Inquiry Results page (p. 82). Observe.

CLEANUP

(All Cleanup suggestions are subject to local laws governing waste from laboratories. The following are suggestions only and may be changed by your instructor. Space has been provided for additional instructions.)

1. The materials you have used are not harmful to the environment in small quantities and may be disposed of by rinsing them down the drain. Any remaining ZnO should be put in the trash.

2. Return all chemicals to their proper places in the lab, and wash and dry your lab bench. Wash your hands before you leave.

3. Additional instructions:

Name ______________________________________ Sec____________

INQUIRY 10: OBSERVATIONS

Set up your Observations page to record all masses and answer all questions. Make observations as you do the Procedure.

Name ______________________________ Sec__________

INQUIRY 10: RESULTS

1. What is the mass of copper in your penny? ____________________

2. From the mass of the penny and the mass of the copper, determine the mass of zinc in the penny.

   ____

3. Calculate the mass of zinc oxide, ZnO, that you produced (the "actual yield").

   ____

4. a. Write the equation for the reaction of Zn with HCl.

   b. Write the equation for the reaction of the product of 4a with Na_2CO_3.

5. Write the equation for the decomposition of zinc carbonate to zinc oxide.

6. What is the mole ratio of the initial zinc to the zinc oxide?

   ____

7. From the starting mass of zinc and the mole ratio of zinc to zinc oxide, calculate the amount of zinc oxide you should have produced (the "theoretical yield"). Show your work.

   ____

8. How did your actual yield differ from the theoretical yield? Explain any difference.

9. Tape your diazo "photo" here. Which is the better sunscreen, the commercial product or the one you made?

Name ______________________________ Sec ____________

INQUIRY 10: FOLLOW-UP QUESTIONS

These are to be done in the laboratory after the Inquiry. You are encouraged to discuss these with your lab partner and your lab instructor.

1. Write the names and list the ingredients of two commercial sunscreen products.

2. What would you assume must be true of the toxicity of ZnO, a compound which is approved for external use on human skin?

3. Look up sunscreen products in your text or in another reference. Several of the chemicals used in sunscreens absorb ultraviolet light. Suggest how ZnO might work as a sunscreen.

4. Write two questions that are inspired by this Inquiry.

   a.

   b.

INQUIRY 11

Qualitative Analysis

The chemist must be able not only to separate substances but to identify the products of the separation as well. Physical methods may be employed to separate two substances, such as the differences in solubilities used to separate solids in Inquiries 4 and 5. We may be interested in *how much* of an element or compound is present (*quantitative analysis*), or we may simply wish to know *what* elements are there (*qualitative analysis*). In this Inquiry, to determine *what* is present we will use differences in chemical properties which will allow us to qualitatively separate and identify three metallic ions, Ag^+, Cu^{2+}, and Fe^{3+}.

Equipment you will need

centrifuge
disposable pipets or droppers
test tubes (several), 100 mm

Chemicals you will use

ammonia, $NH_3(aq)$, 6*M*
copper(II) nitrate, $Cu(NO_3)_2$, 0.1*M*
hydrochloric acid, HCl, 6*M*
iron(III) nitrate, $Fe(NO_3)_3$, 0.1*M*
potassium thiocyanate, KSCN, 0.05*M*
silver nitrate, $AgNO_3$, 0.1*M*
sodium hydroxide, NaOH, 6*M*

BEFORE YOU BEGIN

1. Review definitions of the following.

   a. precipitate:

   b. decanting and decantate:

2. Your instructor will demonstrate the following techniques.

   a safe use of a centrifuge:

   b. shaking a test tube to mix the contents:

SAFETY NOTES

1. Wear approved safety goggles at all times in the lab.

2. Hydrochloric acid, ammonia, and sodium hydroxide are CAUSTIC. Handle them with care. If you get any of these on your skin or clothing, rinse immediately with running water and inform your instructor.

3. All chemicals must be handled carefully and treated with respect. The solids and liquids used in this Inquiry are safe for you to use responsibly.

4. Wash your hands before you leave the lab.

PROCEDURE

Part I: Some Reactions of Each of the Ions

1. Tear out the Inquiry 11 Observations page, and have it ready.

2. Obtain three 100-mm test tubes. Place 10 drops of silver nitrate, $AgNO_3$, in the first one, 10 drops of copper(II) nitrate, $Cu(NO_3)_2$, in the second one, and 10 drops of iron(III) nitrate, $Fe(NO_3)_3$, in the third. (Remember to dispose of Ag compounds as indicated below in Cleanup.)

3. To each of the solutions, add a drop of 6*M* HCl. Record your observations in the table.

4. To the two metal ion solutions which did not form a precipitate in step 3, add 6*M* NaOH dropwise, shaking the test tube between drops, until a heavy precipitate forms in each tube. Record your observations.

5. To each of the same two test tubes, add 6*M* $NH_3(aq)$ dropwise with shaking until a change takes place in one of the test tubes. Record your observations.

6. To the remaining test tube (in which there was no change in step 5), add 6*M* HCl until the precipitate dissolves completely. Then add a drop of potassium thiocyanate, KSCN. What happens?

Part II: Analyzing a Mixture of Knowns

1. At this point it is necessary to construct a flowchart so that you can analyze a solution containing the nitrate salts of the three ions. On the Observations page note the beginning of a flowchart. To complete it, ask yourself some questions based on the observations you obtained in Part I, such as, "Which reagent can I add which will remove one of the metal ions immediately?" And, "What must be added next?"

2. Since all the ions will be in the same test tube you must have some way of separating a precipitate from the ions that are still in solution. Your instructor demonstrated how to use the centrifuge. In your flowchart, note the places in which you will centrifuge to settle the precipitate quickly and decant the solution for further tests.

3. When you have completed your chart, have your instructor approve it before you go on to step 4.

4. Place 10 drops of each of the three metal ion nitrate solutions in a 100-mm test tube. Follow your flowchart and separate this mixture so that the three metal ions are in three different test tubes.

Part III: The "Unknown"

1. When you have finished the separation and identification of the three metal ions, show your results to your instructor and obtain an unknown from her or him. Your unknown may contain one, two, or all three of the ions. Be sure to write the unknown number in the space provided on the Observations page. Follow your flowchart and identify the ions present.

CLEANUP

(All Cleanup suggestions are subject to local laws governing waste from laboratories. The following are suggestions only and may be changed by your instructor. Space has been provided for additional instructions.)

1. Deposit the silver compound in the waste container provided so that it can be recycled.

2. The materials you have used are not harmful to the environment in small quantities and may be disposed of by rinsing them down the drain.

3. Return all chemicals to their proper places in the lab, and wash and dry your lab bench. Wash your hands before you leave.

4. Additional instructions:

Name ________________________________ Sec ____________

INQUIRY 11: OBSERVATIONS

Part I

Write your observations below after addition of the appropriate reagents.

Reagent	Ag^+	Cu^{2+}	Fe^{3+}
HCl			
NaOH			
NH_3			
HCl			
KSCN			

Part II

Complete the following flowchart.

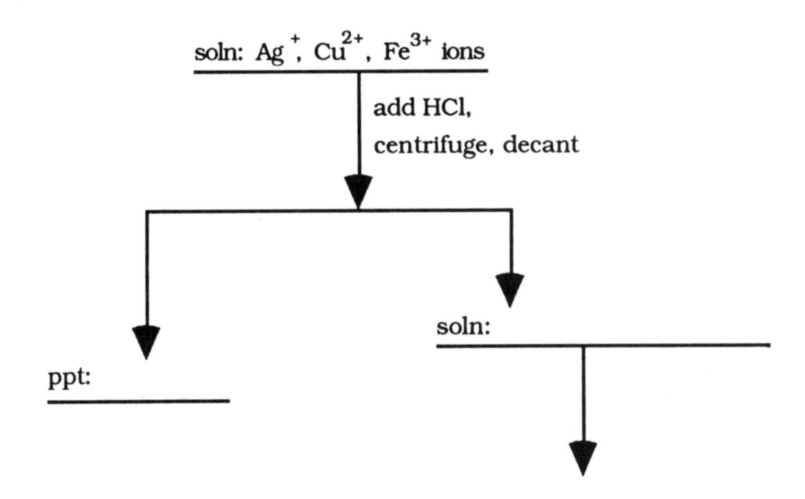

Name ______________________________ Sec__________

INQUIRY 11: OBSERVATIONS (cont'd)

Known solution:

Part III

Unknown solution: Unknown number:__________

The unknown contains the following metal ions:______________________________

Name ______________________________ Sec____________

INQUIRY 11: FOLLOW-UP QUESTIONS

These are to be done in the laboratory after the Inquiry. You are encouraged to discuss these with your lab partner and your lab instructor.

1. Complete balanced equations for each of the following reactions from this Inquiry. If there is no reaction, write "NR."

   a. $AgNO_3$ + HCl → ______________________________

   b. $Cu(NO_3)_2$ + $NaOH$ → ______________________________

   c. $Cu(NO_3)_2$ + HCl → ______________________________

   d. $Fe(OH)_3$ + HCl → ______________________________

   e. $Fe(NO_3)_3$ + $NaOH$ → ______________________________

2. a. Two of the reactions which confirm the presence of metal ions in this Inquiry result in the formation of brightly colored complex ions. In the reaction of the copper(II) ion with ammonia, the brilliant blue complex, $Cu(NH_3)_4^{2+}$, is formed. Write an ionic equation for the formation of this complex. Note that ammonia has no net charge, and that the charge on the complex ion is the same as the charge on the copper(II) ion.

   b. Similarly, the iron(III) ion froms a deep red complex with thiocyanate ion. Since thiocyanate ion is found in this experiment in potassium thiocyanate, KSCN, what is the formula of the thiocyanate ion?

   c. Write an equation for the formation of the iron-thiocyanate complex, assuming the mole ratio of iron(III) to thiocyanate is 1:1.

Name ______________________________ Sec__________

INQUIRY 11 : FOLLOW-UP QUESTIONS (cont'd)

3. You are given solutions of the nitrates of the metallic ions X^{2+}, Y^{+}, and Z^{3+}. Through a series of tests you obtain the following information:

Reagent	X^{2+}	Y^{+}	Z^{3+}
Na_2SO_4	forms white ppt	brown ppt forms	no reaction
$NH_3(aq)$	no reaction	ppt dissolves	no reaction
HCl	no reaction	white ppt forms	no reaction
Na_3PO_4	no reaction	no reaction	white ppt forms

a. Decide how you would start to separate a mixture of these three ions, and then draw a detailed flowchart below.

b. Write an equation for the formation of each of the precipitates in the table.

Unit 3

Additional Chemical Concepts

Where You've Been

Thus far you have learned some central ideas about science in your chemistry course. You have learned, first and foremost, that the work of a chemist begins with observation. You have learned to organize your observations so that you can recognize patterns in them. From these regularities in your observations you have learned to make generalizations and to ask probing questions.

You have begun learning the language of chemistry so that you can communicate your generalizations with others who speak that language. You have used qualitative and quantitative methods to recognize and understand chemical change. You can write names and formulas of compounds, and express chemical change in equations. You can even analyze an unknown. You have delved into the mysteries of chemistry to find that it isn't quite so mysterious after all!

Where We're Going

But there's much more. There are some central concepts that we haven't looked into in detail yet. We have just touched on ideas about acids and bases, but there are unanswered questions, like what makes an acid an acid? Or, how fast does a reaction occur? What makes some reactions, like the burning of gas, very rapid, while others, like the rusting of bridges, take a long time? (Thank goodness!)

Or, why is it that sugar is more soluble in hot tea than in iced tea? Why does a soft drink go "flat" when it is left out in the room? How much acetic acid would you have to put in water to make vinegar? An important question to ask might be this one: Where does the waste from industry and academic research go when it leaves the premises? If we want to get rid of a toxic substance, say a pesticide, can we just add water until the pesticide is so dilute that it is no longer harmful?

Unit 3 is about these questions. Remember, we may find no complete answers. But if we are observant we'll find more good questions to ask along the way.

INQUIRY 12

Acids and Bases

Acids and *bases* are important compounds in our environment. We are familiar with the hazards of acid rain, but we don't often think of the importance of the delicately balanced composition of our blood and other body fluids. We take antacid tablets to neutralize excess acid in our stomachs; we put sulfuric acid in our car batteries; we put baking soda in our biscuits; we may wash clothes with washing soda; we might put vinegar on our salads. Since we generally speak of *acids* as though they were a class of substances, it is important to find out why we classify them together. The same goes for *bases*. What makes them alike? We can classify these materials by using *indicators*, or dyes, many of which are found in nature.

Equipment you will need

droppers
microstirrer
mortar and pestle
vials, several
96-well plate

Suggested household products to test

ammonia cleaning solution
antacid tablet
aspirin tablet
bleach
carbonated drinks
detergents
fruit juice
multivitamin tablet
salad dressing
soaps
vinegar

Other items

paper towels

Chemicals you will use

Laboratory acids and bases, 0.1*M*, in dropper bottles

acetic acid, $HC_2H_3O_2$
ammonia, $NH_3(aq)$
calcium hydroxide, $Ca(OH)_2$
hydrochloric acid, HCl
nitric acid, HNO_3
potassium hydroxide, KOH
sodium hydroxide, NaOH
sulfuric acid, H_2SO_4

Solids

potassium hydrogen phthalate, $KHC_8H_4O_4$ (KHP)
potassium hydrogen tartrate, $KHC_4H_4O_6$ (cream of tartar)
sodium borate, $Na_2B_4O_7 \cdot 10H_2O$ (borax)
sodium carbonate, $Na_2CO_3 \cdot 10H_2O$ (washing soda)
sodium hydrogen carbonate, $NaHCO_3$ (baking soda)

Indicators

bromothymol blue
grape juice
litmus paper, blue and red
methyl red
phenolphthalein
red cabbage

SAFETY NOTES

1. Wear approved safety goggles at all times in the lab.

2. Acids and bases are CAUSTIC in high concentrations. The solutions you are using are very dilute, but if you should spill any of them on your skin or clothing, rinse with running water and inform your instructor.

3. CAUTION: Do not mix household cleaning products.

4. All chemicals must be handled carefully and treated with respect. The solids and liquids used in this Inquiry are safe for you to use responsibly.

5. Wash your hands before you leave the lab.

PROCEDURE

Part I: Laboratory Acids and Bases

1. Set your 96-well plate on a piece of white paper so that you can test each laboratory acid and base with various laboratory indicators: litmus paper, phenolphthalein, bromothymol blue, and methyl red. Start with hydrochloric acid, HCl, and put a few drops in each of 3 wells.

2. Lay a piece of red litmus and a piece of blue litmus paper on a paper towel. Dip the microstirrer in one well of HCl, and touch one piece of litmus. Observe what happens. Rinse the microstirrer, and repeat with another drop on the other piece of litmus. What happens?

3. Now put a drop of phenolphthalein in one HCl well, a drop of bromothymol blue in another, and a drop of methyl red in the third. Record the results.

4. Continue this testing for each acid and base in the list of laboratory acids and bases on the previous page.

Part II: Solids Which May Form Acids or Bases in Solution

1. Put a few crystals or a very small amount (pin head) of one of the solids in each of 3 wells.

2. Add a few drops of water, H_2O, and stir to dissolve the solid. Test with litmus papers and the other indicators as above. Take careful observations.

3. Repeat the same procedure for the other solids.

Part III: Household Products

1. Continue the tests by pouring no more than 1 mL of a liquid household product into a vial and, using your dropper, placing a few drops in each of several wells. (It is not necessary to measure out 1 mL each time you take a sample of a household product. Simply put 1 mL of water in a vial and see how much that is. Get about that amount each time.) The solid products, such as the tablets, can be ground and shared among several pairs of students. Use the mortar and pestle, and clean and dry it between solids.

2. Test each product available with the same indicators as above. Record your observations.

Part IV: Common Foods as Indicators

1. Choose an acid and a base from the list of "Chemicals you will use" list on p. 95, and put a few drops of the acid in one well and of the base in another.

2. Obtain about 1 mL of grape juice, and put a few drops of juice in the acid and a few drops in the base. What happens?

3. Obtain a piece of red cabbage. How might you extract the red coloring agent from the cabbage? Write a short procedure for extracting the dye. After your instructor has approved your extraction method, extract the red dye and use it as an indicator. What happens?

CLEANUP

(All Cleanup suggestions are subject to local laws governing waste from laboratories. The following are suggestions only and may be changed by your instructor. Space has been provided for additional instructions.)

1. The materials you have used are not harmful to the environment in small quantities and may be disposed of by rinsing them down the drain.

2. Return all chemicals to their proper places in the lab, and wash and dry your lab bench. Wash your hands before you leave.

3. Additional instructions:

Name ______________________________________Sec______________

INQUIRY 12: OBSERVATIONS

Parts I and II

Set up your Observations page with a table listing down one side the names and formulas of all the materials tested and listing across the top the indicators. Fill in your observations for each test.

Part III

List the household products you used and your observations.

Part IV

Write your observations of grape juice and red cabbage as indicators.

Name ______________________________ Sec __________

INQUIRY 12: FOLLOW-UP QUESTIONS

These are to be done in the laboratory after the Inquiry. You are encouraged to discuss these with your lab partner and your lab instructor.

1. What generalizations can you make about acids? (One of your generalizations must include what element you think is responsible for acid character.)

2. What generalizations can you make about bases? What element or group of elements seems to be responsible for basic character?

3. Do any of the substances you tested for acid-base character fail to fit the generalizations above?

4. If you wished to dissolve $CaCO_3$, what compound would you choose to do this? Why?

5. Is Na_2CO_3 an acid or a base? Based on your observations of common acids and bases, suggest an explanation for the result you obtained when you tested Na_2CO_3 solution with indicators.

6. Write an equation for the reaction of ammonia, $NH_3(g)$, with water.

7. List five other household products that you would like to test.

INQUIRY 13

Count Those Calories!

For many people the Calorie has become the first consideration in deciding what to eat. "How many Calories does it have?" The number of Calories in food is measured in a special calorimeter that has a chamber in which food can be burned. The chamber sits in a large water bath, the temperature of which can be monitored. The temperature of the water surrounding the combustion chamber rises as the water absorbs the heat released by the combustion. The amount of temperature rise of the water is used to calculate the heat of combustion of a particular food in Calories (the capital "C" indicates kilocalories in food chemistry). We cannot easily duplicate these measurements in a general chemistry lab, but we can do some simple calorimetry which will demonstrate how such measurements are made.

Equipment you will need

balance
beaker, 50 or 100 mL
graduated cylinder, 25 or 50 mL
scoop
thermometer, -10–110°C
watch glass
weighing dish

Other items

cardboard square, 4 x 4 inches, with hole in center
coffee cups (2), Styrofoam

Chemicals you will use

ethanol, C_2H_5OH
hydrochloric acid, HCl, 1.0*M*
sodium hydroxide, NaOH(*s*)

Suggested salts: two salts per pair of students
ammonium chloride, NH_4Cl
ammonium nitrate, NH_4NO_3
copper(II) sulfate, anhyd., $CuSO_4$
copper(II) sulfate pentahydrate, $CuSO_4 \cdot 5H_2O$
potassium bromide, KBr
potassium chloride, KCl
potassium iodide, KI
potassium nitrate, KNO_3
sodium bromide, NaBr
sodium carbonate, Na_2CO_3
sodium chloride, NaCl
sodium iodide, NaI

BEFORE YOU BEGIN

1. Define the following terms.

   a. calorie:

   b. heat:

c. temperature:

d. Celsius:

e. thermal equilibrium:

2. Distinguish between *endo*thermic and *exo*thermic.

SAFETY NOTES

1. Wear approved safety goggles at all times in the lab.

2. Hydrochloric acid, HCl, and sodium hydroxide, NaOH, are CAUSTIC. The solutions you are using are dilute, but if you get any acid or base solution on your skin, rinse with plenty of water and tell your instructor.

3. All chemicals must be handled carefully and treated with respect. The solids and liquids used in this Inquiry are safe for you to use responsibly.

4. Wash your hands before you leave the lab.

PROCEDURE

Part I: Is It endothermic or Exothermic?

1. Tear out the Inquiry 13 Observations page and set up a table for Part I observations. Label the first column "Material to be tested." Label additional columns "Initial temperature" and "Final temperature."

2. Place one coffee cup in the other to produce a well-insulated "coffee-cup calorimeter." Gently push the thermometer through the hole in the center of the cardboard top until it almost, but not quite, touches the bottom of the cup. Place the "cover-thermometer" on the calorimeter, and note that this setup is top-heavy! The thermometer can be used for gentle stirring by moving the cover in a circular motion when solutions are being mixed, but as soon as the final temperature is obtained, lay the cover-thermometer on the bench. It is helpful to work with a partner in all of the following steps. (How will you know when you have reached the maximum or minimum temperature?)

3. Sodium hydroxide, NaOH, is the only solid which must be weighed. The weighed quantity will be used in this part, and the solution you produce will be saved for Part II. Place 20 mL of room temperature water in your calorimeter. While one partner takes and records the initial temperature of the water, the other should weigh out about 0.8 g of NaOH in a weighing dish to the nearest 0.001 g.

4. With the cover-thermometer ready, put the NaOH in the calorimeter and quickly put the cover in place. Gently stir the dissolving solid with the thermometer. (Don't open the calorimeter to see if the solid has dissolved. You will be able to feel the solid as you stir with the thermometer.) Record the final temperature reached, and then remove the cover-thermometer and pour the NaOH solution into a beaker. Cover it with a watch glass, and let the solution return to room temperature while you finish Part I.

5. Your instructor has assigned you and your partner two salts to test. For each of them, use the following procedure. Rinse and dry the calorimeter and the thermometer and again add 20 mL of water. Record the initial temperature of the water, allowing the thermometer and water to reach thermal equilibrium. Place an amount of the salt about the size of a spoonful in the water and stir until dissolved. Record the final temperature reached. Pour the resulting solution down the drain with running water, and rinse and dry the calorimeter and the thermometer. Write your observations for the two salts you tested on the board. Copy all the student observations onto your Observations page.

6. Repeat step 5 using 10 mL of ethanol, C_2H_5OH, in 20 mL of room temperature water.

Part II: Heat of Reaction

1. Obtain 20 mL of 1.0*M* HCl solution and pour it into the calorimeter. Record the initial temperature, t_i, of the solution under a heading of "Part II."

2. Check the temperature of the NaOH solution in the beaker from step 4 of Part I. If it is not the same temperature as the HCl, cool or warm it in a larger beaker of water of the appropriate temperature. When the temperature of the NaOH solution is the same as that of the HCl, pour them together and quickly put the cover-thermometer on. Record the final temperature, t_f.

CLEANUP

(All Cleanup suggestions are subject to local laws governing waste from laboratories. The following are suggestions only and may be changed by your instructor. Space has been provided for additional instructions.)

1. The solutions you used in Part I are not harmful to the environment in small quantities. Dispose of them down the drain.

2. In Part II the products are NaCl and water. Pour this salt solution down the drain.

3. Return all chemicals to their proper places in the lab, and wash and dry your lab bench. Wash your hands before you leave.

4. Additional instructions:

Name ______________________________________ Sec______________

INQUIRY 13: OBSERVATIONS

Name ____Sec______________

INQUIRY 13: RESULTS

Part I

1. Write an equation for solution in water of each of the solutes in Part I. Show the heat as part of the equation, and indicate whether the reaction is endothermic or exothermic. For example:

$$NaCl(s) + \text{heat} \rightarrow Na^+(aq) + Cl^-(aq) \quad \text{(endothermic)}$$

Part II

1. Write the equation for the reaction of aqueous sodium hydroxide with aqueous hydrochloric acid. Is the reaction endothermic or exothermic?

2. To do a rough calculation of the heat of reaction (symbolized by ΔH) of sodium hydroxide with hydrochloric acid, we have to know the change in temperature which the reaction solution underwent and the number of moles of each of the reacting species. To simplify our calculation we will make the assumption that neither the calorimeter nor the surroundings absorbed any of the heat, i.e., that all the heat released by the reaction was used to raise the temperature of the water in which the reaction took place.

   The heat produced by the reaction, q, can be calculated by the equation

$$q = ms\Delta t$$

   where m is the mass of the solution (which, being mostly water, can be calculated from the density of water, 1.0 g/mL), s is the specific heat of water (which is 1 cal/g°C), and Δt is the *change* in temperature, $t_f - t_i$.

Name ______________________________________ Sec______________

INQUIRY 13: RESULTS (cont'd)

a. To calculate q from the equation above, it is necessary to know the total mass of the solution in which the reaction took place. (Hint: What is the mass of 20 mL of water? How much NaOH did you add?)

mass of NaOH solution:__________

What is the mass of the HCl solution you used, assuming that it is so dilute that its density is that of water?

mass of HCl solution:__________

The total mass of solution is therefore ___________.

The temperature change, t_f - t_i, or Δt, is ___________.

Assume the specific heat of the solution you produced is the same as that of water, and calculate q.

b. To make the solution of sodium hydroxide you used 0.8 g of NaOH. Calculate the number of moles of NaOH used.

c. How many moles of NaCl formed?

d. The heat, q, was released when the number of moles of NaCl in your answer to c above were formed in the neutralization reaction. What is the heat of reaction (ΔH) in cal/mole of NaCl formed? By convention, ΔH is positive when a reaction is endothermic and negative when a reaction is exothermic. Be sure the sign of your answer is correct.

e. What is ΔH for this reaction in kcal/mole?

Name ______________________________ Sec__________

INQUIRY 13: FOLLOW-UP QUESTIONS

These are to be done in the laboratory after the Inquiry. You are encouraged to discuss these with your lab partner and your lab instructor.

1. In several of the Inquiries in *Chemistry: The Experience* there is the instruction: "If the solid is slow to dissolve, heat it gently." On the basis of this Inquiry and your understanding of ionic and molecular crystals, explain why that works.

2. What generalizations, if any, can you make about the heats of solution of the solid salts the class tested?

3. Explain the observations you made of anhydrous copper(II) sulfate and of copper(II) sulfate pentahydrate.

4. Write two questions for further research that are suggested to you by the observations you and the class made on the salts.

   a.

   b.

5. Design an experiment to address one of the questions in 4 by writing a procedure that your classmates could follow. (Use a separate sheet of paper.)

INQUIRY 14

Solution Dilution: Answer to Pollution?

Water pollution has become a major concern in the last half of the twentieth century. Beginning with the publication of Rachel Carson's *Silent Spring* in 1962 and spurred on by the first photographs of the Earth from space which gave rise to Earth Day in 1970, our awareness of the fragility of the balance of life on Earth has steadily grown. But with this awareness has come the realization that even though we have allowed industrial waste to pollute the Earth, we can't go backward and give up the chemicals that have made our lives easier and better. The answer is in *personal responsibility*; we are responsible for our waste, whether it be generated in our households, in industry, or in the college chemistry laboratory. One way to minimize the amount of toxic waste in effluent is to recycle toxic materials; another is to destroy the toxic material, producing a nontoxic waste which can go down the drain. It might be reasonable to assume that we could dilute the toxic waste until it is no longer detectable. But that raises some questions. What happens when the test we use is not sensitive enough to measure the concentration of the diluted solution? How many molecules of a toxic substance are too many? We will look at the question of dilution using common laboratory solutions of HCl and NaOH, and we will use the analytical technique of *titration* to help us find some answers.

Equipment you will need

beakers (1 or 2), 100 mL
beakers (2), 250 mL
buret, 25 or 50 mL, and clamp
flasks, Erlenmeyer (2), 125 mL
graduated cylinder, 10 mL
graduated cylinder, 100 mL

Chemicals you will use

hydrochloric acid, HCl, 1.0*M*
phenolphthalein solution
sodium hydroxide, NaOH, 0.1*M*, standardized
water, distilled, H_2O

BEFORE YOU BEGIN

1. Define *effluent*.

2. Today's procedure is called *titration*. Look up the definition of titration in your text or in a dictionary, and write it here.

3. Write the equation for the reaction of hydrochloric acid with a solution of sodium hydroxide. Use your textbook, if necessary.

4. Define *sensitivity* as it relates to measurement.

5. Define *end point*.

6. Your instructor will demonstrate cleaning and filling the buret, and the technique of titration. Take notes here.

SAFETY NOTES

1. Wear approved safety goggles at all times in the lab.

2. Hydrochloric acid, HCl, and sodium hydroxide, NaOH, are CAUSTIC. Be especially careful with the 1*M* solution of HCl. If you spill any acid or base on your skin or clothing, rinse with running water and inform your instructor.

3. All chemicals must be handled carefully and treated with respect. The liquids used in this Inquiry are safe for you to use responsibly.

4. Wash your hands before you leave the lab.

PROCEDURE

Part I: Preparation of the Buret of NaOH

1. Obtain about 60 mL of standardized 0.1*M* NaOH in a 100-mL beaker. Write its actual molarity (from the bottle) in the appropriate blank on the Observations page.

2. Rinse a buret with distilled water and then with 0.1*M* NaOH, as your instructor demonstrated. Fill the buret with 0.1*M* NaOH, taking care to remove bubbles from the tip.

3. Allow NaOH to run into a beaker, until the meniscus is on the scale somewhere between 0 and 1.0 mL. Carefully read the scale to the nearest 0.05 mL, and record this value as the Initial buret reading under "Trial 1" on the Observations page.

Part II: First Dilution and Titration

1. The end point of this titration is indicated by the *faint* permanent pink color of phenolphthalein. Because it is easy to pass the end point and introduce a large error, it is helpful to look at the color you should expect. To see this, place about 60 mL of water in a beaker. Add 1 drop of 0.1*M* NaOH solution and 1 drop of phenolphthalein. The color should be a faint pink.

2. Obtain 10 mL of 1*M* HCl in a 10-mL graduated cylinder. Put 90 mL of distilled water in a 250-mL beaker, and add the HCl solution. Stir to mix. By what factor have you diluted the HCl?

3. Pour 20 mL of the new solution of HCl into one Erlenmeyer flask, and 20 mL into the other. Add 3-4 drops of phenolphthalein indicator to each one. (Label the remaining solution "first dilution" and save it for Part III.)

4. Titrate the solution in one of the flasks by allowing NaOH solution to run into the flask as you mix the resulting solution by swirling the flask gently under the buret tip. As the pink color appears and remains for a longer period of time, slow the addition of NaOH. As you get close to the end point, the color will fade only after several seconds. Slow to dropwise addition, swirling after each drop. One drop may make the difference, so go very slowly at this point. When the faint pink color is permanent, record the final reading of the buret.

5. Refill the buret, record an initial reading under "Trial 2," and titrate the solution in the second flask in the same way, and record the final reading.

Part III: Second Dilution and Titration

1. Refill the buret, and record the initial reading.

2. Put 90 mL of distilled water in a clean, dry 250-mL beaker, and add 10 mL of the first dilution HCl solution. Stir to mix. By how much have you diluted the first dilution solution? By how much have you diluted the stock solution?

3. Pour 20 mL of this second dilution solution of HCl into one Erlenmeyer flask, and 20 mL into the other. Add 3-4 drops of phenolphthalein indicator to each one.

4. Titrate each trial as in Part II. What has happened? Record your observations.

CLEANUP

(All Cleanup suggestions are subject to local laws governing waste from laboratories. The following are suggestions only and may be changed by your instructor. Space has been provided for additional instructions.)

1. The product of the reaction you have carried out here is sodium chloride, NaCl. This dilute solution may be poured down the drain. If you have acid or base remaining, neutralize it and pour it down the drain. Flush with water (dilute it!).

2. Return all chemicals to their proper places in the lab, and wash and dry your lab bench. Wash your hands before you leave.

3. Additional instructions:

Name ____________________ Sec ________

INQUIRY 14: OBSERVATIONS

Part I

Concentration of standardized NaOH: ________

Part II

1st Titration	Trial 1	Trial 2
Final buret reading		
Initial buret reading		
Volume of NaOH		

Other observations:

Part III

2nd Titration	Trial 1	Trial 2
Final buret reading		
Initial buret reading		
Volume of NaOH		

Other observations:

Name ____ Sec ______________

INQUIRY 14: RESULTS

1. Calculate the molarity of the first dilution solution of HCl from the first titration observations, using the molarity and average volume of the NaOH solution and the volume of the HCl solution.

2. Calculate the molarity of the second dilution solution from the second titration observations.

3. In Part III, what has happened to the *sensitivity* of the titration?

4. a. What would be the approximate volume of $0.1M$ NaOH required to titrate a third dilution of the same proportions?

   b. Is it feasible to do a third dilution titration using $0.1M$ NaOH? Why or why not?

5. Suggest a means by which you could increase the sensitivity of the titration so that a third dilution (by 10 again) could be accurately titrated.

Name ______________________________ Sec__________

INQUIRY 14: FOLLOW-UP QUESTIONS

These are to be done in the laboratory after the Inquiry. You are encouraged to discuss these with your lab partner and your lab instructor.

1. The instructions for disposing of hazardous waste from laboratories usually say to destroy the hazardous substance chemically or to precipitate and collect it so that it cannot get into the sanitary waste system and ultimately into the water supply. The final instruction for disposing of the solution remaining after the hazardous substances have been removed is often "Pour down the drain with 50 volumes of water." In other words, if you have a liter of relatively nonhazardous waste (such as NaCl solution from the neutralization reaction in this Inquiry), you should run 50 L of water down the drain with your 1 L of waste. If you had done a 50-volume dilution of the HCl stock solution, would the NaOH solution you used have been able to detect it? What does this say about the effectiveness of a 50-volume dilution for *toxic* materials?

2. One of the maxims of the chemistry lab is that "several small rinses are more effective in cleaning glassware than is one large one." Imagine that you have emptied a 100 mL flask which held an HCl solution. There is still about 1 mL of solution wetting the sides of the flask. Make up an example that will show that three small portions of water, each one added, swirled around, then poured out, will dilute the 1 mL of remaining HCl by a greater factor (thus more effectively cleaning the flask) than will one large portion of water, and that the total of the small portions will be much less volume than the large portion (i.e., that water is conserved).

Name ________________________________ Sec____________

INQUIRY 14: FOLLOW-UP QUESTIONS (cont'd)

3. You have a liter of 1.67*M* pesticide solution. You have been told by your supervisor to add this liter of solution to enough water so that when 1 L of the final solution is tested, there will be only one molecule of pesticide in the liter of solution. What difficulties will you encounter? Write a one-page report to your supervisor detailing the problems you perceive with this approach.

4. Summarize, in a well-reasoned paragraph, what you have learned from this Inquiry and what questions you have. Suggest some areas for further study.

INQUIRY 15

How Much Acetic Acid Is in Vinegar?

Food manufacturers must adhere to stringent rules regarding the chemical composition of the foods we ingest. Quality control chemists in the food industry make certain that products that reach your table conform to Federal guidelines and are safe for your consumption. One such food is *vinegar*. This acidic flavoring agent, used since antiquity, is the natural result of the fermentation of fruit juices. Grape and apple juices ferment to form wine or cider containing ethanol. If the process is allowed to continue, ethanol is oxidized to acetic acid, and vinegar is the product. You will use commercial vinegars in this Inquiry to determine the percentage by weight of acetic acid in vinegar.

Equipment you will need

beakers (2), small
disposable pipets (2), with microtips
test tubes, 100 mm

Chemicals you will use

phenolphthalein solution
sodium hydroxide, NaOH, 0.1*M*, standardized
vinegar, various brands

BEFORE YOU BEGIN

1. Today's procedure is called a *titration.* Look up the definition of titration in your text or in a dictionary, and write it here.

2. Write the equation for the reaction of acetic acid with sodium hydroxide. Use your textbook, if necessary.

3. In your text look up the definition of *weight percent* or *percent by weight.* Write it here.

4. Your instructor will demonstrate how to shake a test tube between your fingers without covering the mouth of the tube.

SAFETY NOTES

1. Wear approved safety goggles at all times in the lab.

2. Sodium hydroxide, NaOH, is CAUSTIC! The solution you are using is very dilute, but rinse any spill on skin or clothing with plenty of water.

3. All chemicals must be handled carefully and treated with respect. The solids and liquids used in this Inquiry are safe for you to use responsibly.

4. Wash your hands before you leave the lab.

PROCEDURE

Note: If you did not do Inquiry 14, turn to page 111, and do Procedure II,1. This will demonstrate the faint pink that signals the end point of this titration.

1. Tear out the Inquiry 15 Observations page, and set it up to take observations.

2. In a small beaker obtain about 5 mL of vinegar. Record the brand name of the vinegar on the Observations page. *Be sure to use the same brand for all three trials.* Obtain a disposable pipet and fill it with vinegar from your beaker. Label the pipet, dry the outside of the pipet, and weigh and record its mass.

3. In another small beaker obtain about 10 mL of standard sodium hydroxide, NaOH. Record the molarity of the NaOH. Fill a disposable pipet with the NaOH. Label the pipet, dry the outside of the pipet, and weigh and record its mass.

4. Put one drop of phenolphthalein solution into a 100-mm test tube. Add 10 drops of vinegar to the same test tube. Gently shaking the test tube back and forth between drops as demonstrated, add the NaOH solution a drop at a time until *one drop* turns the solution a permanent faint pink color.

5. If you add too much NaOH, add 1 or 2 more drops of vinegar, and slowly add NaOH again. After the end point is found, reweigh the disposable pipets and record their masses.

6. Repeat the titration procedure (steps 2-5) twice more using a clean test tube and refilled disposable pipets for each titration. Carefully record all your observations.

CLEANUP

(All Cleanup suggestions are subject to local laws governing waste from laboratories. The following are suggestions only and may be changed by your instructor. Space has been provided for additional instructions.)

1. The solution you have produced by titration is nearly neutral and may be disposed of by rinsing down the drain.

2. Return all chemicals to their proper places in the lab, and wash and dry your lab bench. Wash your hands before you leave.

3. Additional instructions:

Name ______________________________ Sec__________

INQUIRY 15: OBSERVATIONS

Name of vinegar used: __________

% acetic acid on label of bottle of vinegar used: __________

Concentration of sodium hydroxide, NaOH: __________

Set up your Observations page for three sets of weighings.

Name ________________________ Sec ____________

INQUIRY 15: RESULTS

1. Show all calculations on separate paper, and summarize your results below as indicated.

Table 15-1

	Trial 1	Trial 2	Trial 3
mass of vinegar			
mass of NaOH sol'n			
molarity of NaOH			

2. Because the sodium hydroxide and vinegar solutions are so dilute, their densities are essentially that of pure water, or 1.0 g/mL. Using that value, calculate the *volume* of NaOH used in each trial in *liters*, and enter those values in Table 15-2 below. Show your work for Trial 1.

3. From the neutralization equation you wrote in Before You Begin, what is the mole ratio of NaOH to acetic acid, CH_3COOH?

4. Using the concentration of NaOH in moles/L, its volume in L, the mole ratio of NaOH to CH_3COOH, and the molecular weight of CH_3COOH, calculate the mass of CH_3COOH in each trial and enter the figures below in Table 15-2. Show your work for Trial 1.

5. From the mass of CH_3COOH and the mass of vinegar in each trial, calculate the weight % of CH_3COOH in vinegar, and enter the values in the table below. Show your work for Trial 1.

Table 15-2

	Trial 1	Trial 2	Trial 3
volume of NaOH			
mass of CH_3COOH			
wt. % CH_3COOH in vinegar			

Average % CH_3COOH ____________

Name ____ Sec________________

INQUIRY 15: FOLLOW-UP QUESTIONS

These are to be done in the laboratory after the Inquiry. You are encouraged to discuss these with your lab partner and your lab instructor.

1. How does your average % acetic acid in vinegar compare with that on the label of the bottle you used? Assume that the % listed on the label is % by wt.

2. Compare your vinegar with the results obtained by other students for other vinegars.

Vinegar name	Avg % CH_3COOH

3. What generalization can you make regarding the % CH_3COOH in vinegar?

4. Can you think of other foods that contain acids and that could be analyzed by a similar method? Name both the foods and the acid(s) they contain.

5. Benzoic acid, C_6H_5COOH, is a common food additive, used as a preservative. Write the equation for its reaction with sodium hydroxide.

INQUIRY 16

What Affects the Rate of a Reaction?

The rate at which reactions occur is of great importance in our everyday lives, though we are rarely aware that we should be concerned about it. It might seem obvious to us that if we want to start a camp fire it will be a good idea to apply a match to a pile of small twigs, instead of trying to light a log. We know without thinking about it that food spoils faster when left at room temperature than it does in the refrigerator. And if a spot on a carpet is being particularly recalcitrant, we know that a bit more detergent may be the answer. Oxidation of food in our bodies takes place at a controlled rate, but the oxidation of hydrogen that produced the *Hindenburg* and *Challenger* disasters was oxidation occurring at a rate that we term "explosive." In Inquiry 7 we compared the rates of reaction of an acid with various metals. In this Inquiry we will look at the ways in which changing reaction parameters such as concentration of the reacting species or the temperature of the system can affect the rate of a reaction, and we will briefly examine catalysts and their importance in our lives.

Equipment you will need

beakers (2), 100 mL
Bunsen burner
droppers and disposable pipet
forceps, plastic
graduated cylinders, 10 and 50 mL
ring stand and ring
test tubes (4), 150 mm
test tubes (6), 100 mm
thermometer, -10–110°C
24-well plate or small test tubes
wire gauze

Other items

hydrogen peroxide, commercial
ice
matches (2)
starch, soluble
steel wool
turnip, very small piece
wood splint

Chemicals you will use

Solids
- calcium, Ca (turning)
- magnesium, Mg (ribbon)
- manganese dioxide, MnO_2
- sodium metal (instructor only)
- zinc, Zn, granulated
- zinc, Zn, square, approximately 5 x 5 mm

Solutions
- hydrochloric acid, HCl, 3*M* (dropper bottle)
- hydrogen peroxide, H_2O_2, 3% by vol
- phenolphthalein solution
- potassium iodate, KIO_3, 0.02 *M*
- sodium hydroxide, NaOH, 0.1*M* (dropper bottle)
- starch-bisulfite solution

BEFORE YOU BEGIN

1 Define the following terms

a. catalyst:

b. rate of reaction:

c. concentration:

d. enzyme:

e. inhibitor:

2. Look up the *Hindenburg* distaster in your text or in an encyclopedia, and, on a separate sheet of paper, describe the *Hindenburg* and explain what happened to it.

SAFETY NOTES

1. Wear approved safety goggles at all times in the lab.

2. All chemicals must be handled carefully and treated with respect. The solids and liquids used in this Inquiry are safe for you to use responsibly.

3. Wash your hands before you leave the lab.

PROCEDURE

Part I: Some General Observations

1. Tear out the Inquiry 16 Observations page and have it ready to take observations in each step of the Procedure.

2. Your instructor will demonstrate the reaction of sodium metal, Na, with water, in the presence of phenolphthalein. Record your observations.

3. Place 3 or 4 drops of 0.1*M* sodium hydroxide, NaOH, in a well of your 24-well plate. Add 1 drop of phenolphthalein indicator solution. How rapidly does the base react with the indicator? Record your observations.

4. Put water in a well of your 24-well plate to a depth of 2-3 mm, and place the plate on a piece of white paper. Add a drop of phenolphthalein. Using steel wool, polish a 1-cm piece of magnesium ribbon, Mg, and drop it into the water. What happens? Let it sit and observe again a few minutes later. Compare your observations with those of the instructor's demonstration.

5. Again put a small amount of water in a well, add a drop of phenolphthalein, and this time add a small piece of calcium metal, Ca. Record your observations. Again compare your observations with those of the instructor's demonstration.

6. Obtain 5 mL of 0.02*M* potassium iodate solution, KIO_3, in a 150-mm test tube. In another test tube obtain 5 mL of starch-

bisulfite solution. Pour the two solutions together. What happens? Continue watching. Does anything happen?

7. If you did not do Inquiry 7 then turn to (p. 54) and carry out Procedure II,A: Reactions with acid (steps 1 and 2). Record your observations. If you did do Inquiry 7, copy your observations from Part II,A onto your present Observations page.

Part II: Varying Concentration

A. Solution concentration

1. Obtain 15 mL of 0.02*M* potassium iodate solution, KIO_3. Put 10 mL of KIO_3 solution into one 150-mm test tube (label it 1A), and 5 mL into another (2A). To the second tube add 5 mL of distilled water and stir, so that the volumes of solution in the two tubes are equal.

2. Obtain 20 mL of starch-bisulfite solution, and divide it equally into two 150-mm test tubes, which you label 1B and 2B.

3. Set up your Observations page to time two reactions. "Time zero" is the moment the solutions contact each other. While your partner times, pour solutions 1A and 1B together. Then do the same for 2A and 2B.

B. Solid surface area

1. Place a small square of Zn in a well of your 24-well plate. Add a few drops of 3*M* HCl. Observe.

2. Place a small pile of granulated zinc (about the size of a match head) in a well. Add a few drops of 3*M* HCl.

3. Place your wire gauze on the bench. Put a small pile of soluble starch about the size of two peas on the ceramic center of the gauze. Hold a lighted match to the starch. Observe. Save the pile of starch.

4. Using a plastic pipet, pick up the starch from the pile on the gauze. Light a burner, and, holding the pipet close to the flame, compress the bulb to blow the sample into the flame. What happens?

Part III: Varying Temperature

1. Record room temperature. Put about 75 mL of water in a 100-mL beaker, heat it gently to raise the temperature about 10°C above room temperature, and then turn off the burner. While the water is heating, set up a cold bath by putting a few small pieces of ice in about 75 mL of tap water in another 100-mL beaker. Lower the water temperature to about 10°C below room temperature.

2. Obtain 9 mL of 0.02*M* potassium iodate solution, KIO_3, and 9 mL of starch-bisulfite solution. Set up six 100-mm test tubes. Divide the 9 mL of KIO_3 solution equally among the first three test tubes. Divide the 9 mL of starch-bisulfite solution equally among the last three tubes.

3. Put one tube of each reactant in step 2 into the warm water bath, and one tube of each reactant into the cold water bath. Retain the third set for the room temperature determination. Allow the solutions about 5 minutes to attain the temperatures of the baths.

4. Prepare to time the reactions at the three temperatures and to record your observations. Start with the room temperature reactants. "Time zero" is the moment the solutions contact each other. Pour one reactant into the other. Observe and time the reaction. Do the same for the other two sets of reactants.

Part IV: Catalysts

1. Although the hydrogen peroxide you will use in this part is not the commercial one (commercial H_2O_2 may contain a chemical which inhibits the decomposition of the peroxide), read the label of the commercial hydrogen peroxide and write down the ingredients, the color of the bottle, and any directions for storing it.

2. Locate the laboratory reagent hydrogen peroxide. Obtain about 6 mL of this 3% H_2O_2 solution, and dispense it equally into three 100-mm test tubes.

3. Obtain a wood splint, and light a Bunsen burner. Leaving the first test tube as a control, add a small amount (size of a match head) of MnO_2 to the second tube. To the third tube, add a small piece of turnip.

4. Light the splint and blow it out gently so that it continues to glow. Insert it quickly into the top of the first tube. Observe. Relight, blow the flame out again, and insert the glowing splint into the top of the second tube. Repeat with the third tube. Observe.

CLEANUP

(All Cleanup suggestions are subject to local laws governing waste from laboratories. The following are suggestions only and may be changed by your instructor. Space has been provided for additional instructions.)

1. The materials you have used are not harmful to the environment in small quantities. The solids should be picked out of the well plate with plastic forceps and placed in a solid waste container, or put directly in the trash (your instructor will tell you which). Liquid waste may be disposed of by rinsing it down the drain.

2. Return all chemicals to their proper places in the lab, and wash and dry your lab bench. Wash your hands before you leave.

3. Additional instructions:

Name ______________________________ Sec____________

INQUIRY 16: OBSERVATIONS

Name ______________________________ Sec____________

INQUIRY 16: RESULTS

Look at your observations carefully. Organize them, and draw conclusions about the effects of concentration, temperature, and catalysts on the rates of chemical reactions.

Name ______________________________ Sec____________

INQUIRY 16: FOLLOW-UP QUESTIONS

These are to be done in the laboratory after the Inquiry. You are encouraged to discuss these with your lab partner and your lab instructor.

1. Iron, in the form of the alloy *steel*, is our most important structural metal, but the rusting of iron costs billions of dollars annually in replacement steel. Rust is formed when iron reacts with air in the presence of water.

   a. What can you say about the rate at which iron rusts? (Think about the rusting of a tool left outdoors in relation to the reactions you have observed today.)

   b. How does painting an iron surface retard the rate of corrosion (rusting)?

2. a. Draw a cube, and label its dimensions 2 x 2 x 2 inches. Determine the total surface area of the cube.

   b. Now divide each face of the cube into four equal sections. If you were to take the cube apart on the lines you have just drawn, how many cubes would you now have?

   c. What are the dimensions of the new cubes?

   d. What is the total surface area of all the cubes now?

Name ________________________________Sec____________

INQUIRY 16: FOLLOW-UP QUESTIONS (cont'd)

e. In wheat-growing areas, such as the plains regions of the United States and Canada, harvested wheat is stored in tall grain elevators that are visible for miles in the flat prairie. The wheat is dumped from trucks into a lower area of the elevator and then moved up the elevators into storage areas. The combustible dust produced as the wheat grains rub together has caused grain elevator explosions in the past.* Use the cube analogy and the starch burning experiment in the Inquiry to explain why wheat dust dispersed in the air in a grain elevator can be explosive if sparked, whereas wheat grains would not be.

3. a. What is the name of the enzyme in the turnip that caused the reaction you observed?

   b What is the "lock and key" model of enzymes?

4. According to the kinetic molecular theory, reactions occur when molecules collide with sufficient energy and correct geometry to rearrange and form products. Explain why increasing the concentration of one of the species in a reaction might increase the rate of product formation.

---

* Recently the U. S. Department of Agriculture approved spraying the wheat with a light mist of food grade mineral oil which greatly reduces the problem of wheat dust in the elevators. The oil has the added advantage of helping maintain moisture levels in the wheat.

INQUIRY 17

It's a Gas!

Carbon dioxide, CO_2, is an excellent example of a compound that is both essential for plant life and a waste product of all living systems. Without it, life as we know it on Earth would not exist, but too much CO_2 in the atmosphere could potentially make the Earth uninhabitable. As you know, a very delicate balance must be maintained if all living systems are to function optimally. Gaseous CO_2 is used in this Inquiry to study some properties of gases, and because it has some properties that make it unique among common gases, it gives us an opportunity to understand why the ecological balance is so fragile.

Equipment you will need

beakers, various
Bunsen burner
graduated cylinder, 50 mL
ring stand and ring
stirring rod
stoppers (4), 2 hole, size-00
test tubes (4), 100 mm
thermometer, -10–110°C
tongs or forceps
wire gauze

Chemicals you will use

bromothymol blue indicator
hydrochloric acid, HCl, 0.1*M*, in dropper bottle
calcium hydroxide, $Ca(OH)_2$, sat'd (limewater)
sodium hydroxide, NaOH, 0.1*M*, in dropper bottle

Other items

antacid tablets (2), effervescent
canned soft drink (kept on ice)
dry ice
ice

Equipment and chemicals for instructor's demonstration

balloons (3), light-colored
bucket or very large evaporating dish
gloves
lecture bottles of $He(g)$, $CO_2(g)$, and $O_2(g)$
nitrogen, liquid, in a Dewar flask
string for balloons

BEFORE YOU BEGIN

1. Define the following terms.

   a. gas:

   b. effervescent:

c. indicator:

d. dry ice:

SAFETY NOTES

1. Wear approved safety goggles at all times in the lab.

2. Liquid nitrogen can freeze skin. Handle it very carefully in a wrapped Dewar flask. It must not be poured into the sink or down the drain. To dispose of it, simply let it evaporate.

3. Dry ice can also freeze skin. Handle it with tongs. Do not put it into the sink. To dispose of it, let it sublime.

4. All chemicals must be handled carefully and treated with respect. The other solids and liquids used in this Inquiry are safe for you to use responsibly.

5. Wash your hands before you leave the lab.

PROCEDURE

Tear out the Inquiry 17 Observations page, and set it up to take careful observations at each step of the Procedure, including the instructor's demonstration.

Part I: Instructor's Demonstrations

The following three steps should be carried out by the instructor, with students taking observations.

1. Fill one balloon with $CO_2(g)$, one with He(g), and the third with $O_2(g)$. Knot the mouth of each balloon, and tie a string around the knot. Which gas is heavier than air?

2. In this step students should be prepared to compare the times required for each balloon to return to its original condition. Hold the He balloon over a very large evaporating dish or bucket, and, wearing gloves to protect your hands, pour liquid nitrogen over the He balloon. What happens? Now pour liquid nitrogen over the O_2 balloon. Describe what happens. Is the response of this balloon the same as the previous one? Finally, pour liquid nitrogen over the CO_2 balloon. How does it compare to the He and O_2 balloons?

3. Have the students pass the CO_2 balloon around while it is warming. Each student should hold the balloon up by the knot and look through the underside of the balloon against the light. Shake the balloon. What is in the balloon now?

Part II: Preparing to Generate and Test CO_2

1. Obtain four 100-mm test tubes and 1-hole stoppers to fit them.

2. Prepare a hot water bath by heating about 500 mL of water in a 600-mL beaker. Stand a thermometer in the beaker, and go on to step 3, but check the temperature of the water frequently. When the temperature reaches about 60°C, turn the burner off. Continue to monitor the temperature periodically, and when it drops below 50°C, heat to about 60°C again.

3. Set up an ice bath by filling a 250-mL beaker with ice and adding water. Pour about 30 mL of this cold water (holding back the ice) into a 50-mL beaker, and set the 50-mL beaker in the ice bath to keep the water cold. Put 1-hole stoppers in three of the 100-mm test tubes, and place the tubes, stopper up, in the 250-mL ice bath to cool them.

4. Obtain 20 mL of lime water [saturated calcium hydroxide, $Ca(OH)_2$]. Put about 10 mL of the limewater in a clean 50- or 100-mL beaker, and in another beaker place10 mL of water + 3 drops of bromothymol blue indicator. Using tongs or forceps, obtain two little pieces of dry ice. Place a piece of dry ice in the limewater and another piece in the water + indicator. What happens? If you don't remember what color bromothymol blue is in acidic or basic solution (Inquiry 12: Acids and Bases), test a drop of indicator in a few drops each of HCl and NaOH.

Part III: Generating and Observing CO_2

1. Break an effervescent antacid tablet in half and put one half in the cold water in the 50-mL beaker. Observe. If small pieces of the tablet remain after a minute or so, stir to break them up. Wrap up the remaining half carefully.

2. This step and the next must be done quickly, so be ready to time the generation of the CO_2 in seconds. Also, have some limewater and bromothymol blue at hand. Fill two of the cold 100-mm test tubes to the rim with the cold antacid solution and stopper firmly with 1-hole stoppers. Invert the test tubes (why doesn't the solution drain out?) and *immediately* place them stopper down, in the hot water bath. What happens? How long does it take? What is in the test tubes now?

3. As soon as the reaction is complete, remove the test tubes and place them upright. (Why?) Quickly remove the stopper from one and pour about 3 mL of limewater into the tube, restopper, and, placing your thumb over the hole in the stopper, shake the tube to mix limewater and gas. Into the other test tube, quickly place 3 mL of water and a drop of bromothymol blue. Shake this tube as well. Observe both tubes.

Part IV: Solubility of CO_2 in Water at Room Temperature and at 10°C Above Room Temperature

1. You have already examined a cold solution of antacid in water. In this part you will make a room temperature solution and a warm water solution.

2. Put about 30 mL of room temperature water in a small beaker. Add half an effervescent antacid tablet, and stir to break up small bits. Quickly pour the solution into a room temperature 100-mm

test tube to the rim, and stopper as before with a 1-hole stopper. Invert the tube in the hot water bath. Observe the time it takes to complete gas evolution. How much gas evolves?

3. You will need a warm water bath and a 100-mm test tube at the same temperature as the bath. Make the warm water bath by using water from the hot bath. Using beaker tongs, pour about 40 mL of hot water into a 100 mL beaker, and add 40 mL of tap water. Stand a stoppered, empty test tube in the warm bath for a minute or two so that its temperature will be about that of the water. Then pour about half of the warm water into a 50-mL beaker, and drop half an effervescent tablet into the water. When the reaction stops, fill the test tube with the warm water solution, stopper with the 1-hole stopper as before, and invert it in the hot water bath. Observe.

Part V: Reversing What the Soft Drink Company Did

Be prepared to time the gas release in the following, and choose one of the confirmation tests for CO_2 that you used in Part II,4 (does one of the tests give a more obvious result?). Fill the third cold empty test tube with cold soft drink. (Be sure to return the soft drink can to the ice to keep it cold for the next person.) Stopper the test tube with a 1-hole stopper, and quickly invert it in the hot water. Observe the time gas evolution takes. If some soft drink remains after gas evolution, very quickly pour out the liquid and restopper. Confirm the presence of carbon dioxide with the test you have chosen.

CLEANUP

(All Cleanup suggestions are subject to local laws governing waste from laboratories. The following are suggestions only and may be changed by your instructor. Space has been provided for additional instructions.)

1. Dispose of liquid nitrogen by letting it evaporate. Allow dry ice to sublime.

2. The solutions you have used are not harmful to the environment in small quantities and may be disposed of by rinsing them down the drain.

3. Save the aluminum can from the soft drink for Inquiry 24.

4. Return all chemicals to their proper places in the lab, and wash and dry your lab bench. Wash your hands before you leave.

5. Additional instructions:

Name ______________________________ Sec__________

INQUIRY 17: OBSERVATIONS

Name ______________________________ Sec ____________

INQUIRY 17: RESULTS

1. If a large quantity of gas evolved quickly from the cold solution when the temperature of the solution was suddenly raised in the hot water bath, what does this say about the solubility of carbon dioxide in cold water? In other words, when you dissolved the tablet in the cold water, do you think much carbon dioxide went into solution? Explain.

2. Write a generalization that can be made from your observations about the solubility of carbon dioxide gas in water at various temperatures.

3. Write the equation for the reaction of calcium hydroxide with carbon dioxide.

4. Hydrochloric acid, HCl, is added to the precipitate formed in 3 above. Write the equation for the reaction that occurs.

Name ______________________________ Sec ____________

INQUIRY 17: FOLLOW-UP QUESTIONS

These are to be done in the laboratory after the Inquiry. You are encouraged to discuss these with your lab partner and your lab instructor.

1. The boiling point of liquid nitrogen is -196°C. The melting point of carbon dioxide is -56.5°C. When you look through the CO_2 balloon which has been cooled by liquid nitrogen and which is warming back to room temperature, why do you not see a liquid phase? To answer this you may have to look up carbon dioxide in your text book or in a chemical handbook.

2. You made observations with a single gas, CO_2. You know from past study that fish take dissolved O_2 from water by filtering the water over their blood-enriched gills.

   a. Using a chemical handbook, look up the solubilities in water of O_2 and CO_2. How do they compare?

   b. What do you suppose would happen if you were to heat water which contained dissolved oxygen? Give evidence from daily life that supports your answer.

   c. What generalization can you now make about the solubility of gases in water and the temperature of the water?

   d. Think about what occurs on the molecular level, and write an explanation for the generalization you made in c.

3. Although the soft drink conformed to the same solubility rule that was established for gases dissolved in liquids at different temperatures, CO_2 was dissolved in the soft drink in the first place by employing a different property of gases. Look up this property in a textbook, or call a local bottling plant or the 800-number of a national office of a soft-drink company and ask how carbonation is done. Write a summary of your findings.

Name ______________________________ Sec __________

INQUIRY 17: FOLLOW-UP QUESTIONS (cont'd)

4. The reaction that produces CO_2 from an effervescent antacid tablet is quite simple. When the tablet dissolves in water, bicarbonate ions from sodium bicarbonate (baking soda) react with hydrogen ions from the solid acid (citric acid) contained in the tablet.

   The net reaction is

$$HCO_3^-(aq) + H^+(aq) \rightarrow H_2O + CO_2(g)$$

   a. If one mole of citric acid donates one mole of H^+ ions, how many moles of carbon dioxide can be produced?

   b. An effervescent antacid tablet contains 0.8 g of citric acid (molecular weight 192 g/mol). If one mole of carbon dioxide gas occupies 22.4 L at STP (standard temperature, 0°C, and standard pressure, 1 atmosphere), how many liters of carbon dioxide could be produced by the citric acid in one effervescent antacid tablet at STP?

Special Project

Your instructor may assign this Follow-up Special Project.

Imagine that each day a new industry in your city pumps a large quantity of water from an area lake through a cooling system that removes heat from a manu-facturing process. The water is returned to the lake. There are no hazardous chemicals introduced into the water in this process. The only change that can be monitored is that the surface water in the lake now averages 5°C warmer than it has in the past few years. You have been invited to a meeting at which you will argue that this temperature change could have a deleterious effect on the ecology of the lake. In light of this Inquiry, and keeping in mind the interrelatedness of life processes, discuss this problem with your classmates, and write a well-reasoned argument. [Some things to consider: 1) Look up the density of water at different temperatures. 2) What are the needs of aquatic life near the bottom of the lake?]

Unit 4

Chemistry Today

Chemistry is more than a study of molecules and atoms, of principles of reaction. Once you have some understanding of how reactions occur, then you can begin applying those principles to solve problems. For example, for thousands of years people have used natural dyes to color their world. Metal oxides, such as rust-red Fe_2O_3, color the hieroglyphic paintings in Egyptian tombs, where clothing dyed with vegetable dyes has also been found; insect bodies, such as the scarlet colored cochineal, provided dyes for ancient Mexico; in the ancient Near East, a Mediterranean mollusk secreted a fluid which produced the royal "Tyrian purple" dye which the Phoenicians sold to those wealthy enough to afford it, namely, the Roman emperors. But because thousands of mollusks have to be harvested and processed to provide a few grams of Tyrian purple dye, we can see from a modern perspective that it makes sense to use chemistry to determine the molecular structure of the dye and to synthesize it. In 1856 the first synthetic dye was made, and the dye industry was born.

Not only do we want to wear colorful clothes, we like to eat colorful food. Grape and orange soda would be colorless liquids without food dyes. These dyes, however, pose an additional problem—they must be safe to eat! Because of this, an interesting circle of events has occurred in the food dye industry. One of the earliest food dyes used on a commercial scale in the late 1800s was the vegetable dye, annatto, which was added to margarine. As the production of synthetic dyes became less expensive it was logical to produce synthetic food dyes. The agency which tested and listed these as safe to consume was the Food and Drug Administration (FDA). In recent years, the FDA has had to "delist" several dyes due to their potential carcinogenicity. This fate befell F, D, & C (Food, Drug, and Cosmetic) Red #2. So, in search of safe dyes, not only does the food dye industry include synthetic dye research, but its search has come back to the beginning as well — the search for natural dyes to color food.

One thing that we must be careful about, however, is assuming that "natural" means better or safer. We may decide to grow food without adding synthetic pesticides, but we must be aware that there are many chemicals in nature that are toxic to humans. For example, the potato is a member of the family Solanaceae, which includes tomatoes, peppers, tobacco, and one of the most famous deadly poisonous plants, belladonna, or deadly nightshade. The potato is perfectly harmless and even a valuable source of vitamins until it is exposed to light, which turns potato skin green. The green part contains a toxin that should not be eaten. Tobacco contains the toxic chemical nicotine, which can be used as a natural insecticide. And in a marvelous twist of nature, deadly nightshade, which can cause death if eaten, contains the sedatives atropine and scopolamine which are extracted and used as ingredients in medicines!

So chemistry today is a complex subject addressing complex issues. Whether the question is one of providing energy resources safely and inexpensively, of providing safe, nutritious food for our tables, of providing new materials for packaging and textiles, or even one of satisfying a need to express individuality (and sometimes conformity) in the colorful clothing we wear, chemistry is involved. In this Unit we will look at some of the ways chemistry pervades our lives.

INQUIRY 18

"Contains Artificial Flavors and Colors"

Many of the foods we buy, from bright red canned cherries to "junk foods" such as soft drinks and chips, have a statement in the ingredient section of the label that alerts one to the artificial colors and flavors that are used to make the food appealing. Manufacturers even put artificial colors in pet food to attract human purchasers. In this Inquiry you will examine some of the colors used to create brightly colored foods, like diet grape soft drink and lime gelatin dessert, using a separations technique known as *chromatography*. To examine some flavorings, you will make one of the compounds used to flavor items like chewing gum and mints, and you will sample the odors of some other familiar flavoring agents.

Equipment you will need

beaker, 250 mL
Bunsen burner
capillary tubes, 1.8 mm
evaporating dish(es)
flask, Erlenmeyer, 250 mL, or a jar
graduated cylinder, 50 mL
ring stand and ring
spatula
stopper to fit flask
wire gauze

Chemicals you will use

Laboratory chemicals
- ethanol, C_2H_5OH
- methanol, CH_3OH
- rubbing alcohol (70% isopropyl alcohol, C_3H_7OH, 30% water)
- salicylic acid, $C_7H_6O_3$
- sulfuric acid, H_2SO_4, conc.
- water, distilled, H_2O

Others
- diet grape soft drink
- diet lime gelatin dessert
- food dyes (red, blue, yellow)

Other items

chromatography paper
"flavor vials" containing cotton saturated with flavoring compounds, such as benzaldehyde, cinnamaldehyde, ethyl acetate, isoamyl acetate, menthol, and vanillin

BEFORE YOU BEGIN

1. Look at any soft drink can, and list the ingredients on the label.

2. Define the following terms.

   a. chromatography:

   b. adsorption:

   c. flavoring:

   d. ester:

   e. aldehyde:

   f. alcohol:

3. Your instructor will demonstrate the following.

   a. how to spot chromatography paper with a sample, using a capillary tube:

   b. how to smell a chemical safely:

SAFETY NOTES

1. Wear approved safety goggles at all times in the lab.

2. In this Inquiry you will use your sense of smell to identify some flavorings. Smelling a substance must always be done carefully, even when smelling a chemical that is safe (in small quantities) for human consumption. To detect the odor, WAFT the vapors toward your nose.

3. Some compounds, especially esters and other flavoring compounds, can sometimes cause ALLERGIC REACTIONS. If you have a history of allergies, inform your instructor.

4. Ethanol, C_2H_5OH, is FLAMMABLE. Be sure that it is used a safe distance from flame.

5. Methanol, CH_3OH, is FLAMMABLE and TOXIC. Use it at a safe distance from a flame, and avoid breathing the vapor.

6. Concentrated sulfuric acid, H_2SO_4, is very CAUSTIC. If you spill any on skin or clothing, rinse with running water *immediately* and inform your instructor.

7. All chemicals must be handled carefully and treated with respect. The solids and liquids used in this Inquiry are safe for you to use responsibly.

8. Wash your hands before you leave the lab.

PROCEDURE

Preliminary Procedure

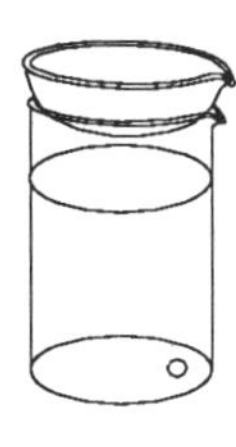

Since ethanol and methanol are flammable, it is important to divide your bench into a heating area (where the Bunsen burner will be used) and a work area. You will need a steam bath or hot water bath in several procedures below, so before starting, set it up by determining on which of your beakers your evaporating dish will sit comfortably. Fill that beaker 3/4 full of water, add a boiling bead, and begin heating the water to boiling. Go on to Part I. Turn off the burner when the water boils.

Part I: Paper Chromatography and Food Colors

1. Tear out the Inquiry 18 Observations page, and have it ready to take organized observations.

2. Mix one drop each of blue, red, and yellow food colors in a vial. Add 1 mL of ethanol.

3. Set up a chromatography chamber in a 250-mL Erlenmeyer flask (or as demonstrated by your instructor). Put 20 mL of rubbing alcohol into the flask fitted with a rubber stopper.

4. Obtain a piece of chromatography paper a little longer than the height of the flask. At about 1 cm from one end of the paper, draw a *pencil* line (not ink) across the paper. This provides a place for you to spot the paper with the mixture of food dyes.

5. Stick a capillary tube into the mixture of food dye so that a sample of dye rises in the tube. Touch the chromatography paper with the sample on the pencil line you made, allowing a small amount of dye to make a spot on the paper. Let the spot dry.

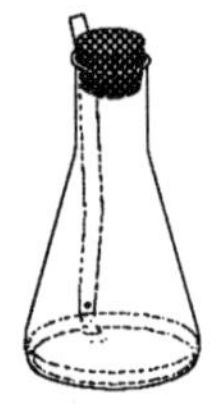

6. Hang the chromatography paper in the flask, using the stopper to hold it so that the dye spot is about 0.5 cm from the surface of the rubbing alcohol and the end of the paper below the spot is just under the surface. Leave the chromatogram for about 30 minutes and go on to Part II.

7. When the dyes are separated, remove the paper from the flask and staple it to your Observations page. Label it and describe what you see.

Part II: Extracting the Dyes from Foods

Your instructor will assign you to do either II,A or II,B.

A. Diet grape soft drink

1. Write down the ingredients on the label of the can of diet grape soft drink.

2. Because the dye concentration in the drink is very small, the dye must be concentrated by evaporating most of the liquid. Pour about 3 mL of grape drink into an evaporating dish, and set the dish on the steam bath that you prepared earlier.

3. When the water has evaporated, add 4 drops of ethanol and 1 drop of water. Using a spatula, scrape the dye until it is dissolved in the solvent mixture. Note the color.

4. Prepare a fresh chromatography paper, drawing a pencil line as you did in Part I. Write "grape drink" on the other end of the paper. Using a clean capillary tube, spot the paper with the dye on the line as before. Let the first spot dry completely. To build up enough dye to get a good chromatogram it is necessary to continue adding dye to the same spot, allowing the spot to dry after each addition. Four applications of dye should suffice. Dry the final application, and then put the paper into the flask with the other chromatogram. Go on to Part III.

B. Diet lime gelatin dessert

1. Write down the ingredients on the label of the box of diet lime gelatin.

2. Pour diet lime gelatin crystals into a 100-mm test tube to a depth of about 1 cm. Add 2 mL of ethanol and a few drops of distilled water. Stir to dissolve the dye, and pour the resulting solution into a clean evaporating dish. Be sure that the Bunsen burner is off, and set the evaporating dish on the steam bath.

3. When the solvent has evaporated, add 4 drops of ethanol and 1 drop of water. Using a spatula, scrape the dye until it is dissolved in the solvent mixture. Note the color.

4. Prepare a fresh chromatography paper, by drawing a pencil line as you did in Part I. Write "lime gelatin" on the other end of

the paper. Using a clean capillary tube, spot the paper with the dye on the line as before. Let the first spot dry completely. To build up enough dye to get a good chromatogram it is necessary to continue adding dye to the same spot, allowing the spot to dry after each addition. Four applications of dye should suffice. Dry the final application, and then put the paper into the flask with the other chromatogram. Go on to Part III.

Part III: Flavorings

A. Synthesis of a familiar flavor

1. Reheat the steam bath to boiling.

2. While the water is heating, place a pea-sized amount of salicylic acid, $C_7H_6O_3$, in a clean evaporating dish, and add 3 mL of methanol, CH_3OH. *Carefully* add 10 drops of concentrated sulfuric acid. When the water is boiling vigorously, turn off the flame, and place the evaporating dish on the steam bath for 5 minutes.

3. Waft the vapors from the dish to your nose carefully. What have you made? Dispose of this mixture by pouring it into a beaker of water. Set it aside for Cleanup. It still contains sulfuric acid. (Why?) What is floating on the water?

B. A potpourri of flavorings

Your instructor has set out some corked vials, each of which contains a piece of cotton soaked with a common flavoring compound. Your observations of the vials should include the flavor name (such as "spearmint") and the chemical name from the label.

CLEANUP

(All Cleanup suggestions are subject to local laws governing waste from laboratories. The following are suggestions only and may be changed by your instructor. Space has been provided for additional instructions.)

1. Add Na_2CO_3 to the beaker of waste from Part III until the fizzing stops. Pour the solution down the drain with water.

2. The materials you have used are not harmful to the environment in small quantities and may be disposed of by rinsing them down the drain.

3. Return all chemicals to their proper places in the lab, and wash and dry your lab bench. Wash your hands before you leave.

4. Additional instructions:

Name ______________________________________ Sec______________

INQUIRY 18: OBSERVATIONS

Organize your observations of the various parts of the Procedure. Attach and interpret your chromatograms.

Name ______________________________ Sec____________

INQUIRY 18: FOLLOW-UP QUESTIONS

These are to be done in the laboratory after the Inquiry. You are encouraged to discuss these with your lab partner and your lab instructor.

1. a. What is the common name of the flavoring that you produced?

   b. What is its chemical name?

   c. What kind of compound is this flavoring? Write its formula.

2. Why was a "diet" food specified in each of the dye extractions that you did?

3. Use your textbook to help you identify three classes of organic compounds used in this Inquiry, and give two examples of each one from the Inquiry.

Special Projects

Your instructor may assign one of these Follow-up Special Projects.

1. Write a report on "scratch and sniff" samples used in advertising. How do they work? Why are they controversial?

2. Write a report on the controversy surrounding some food dyes, such as F, D, & C Red #2.

3. Until food flavorings were synthesized commercially in the twentieth century, people depended on natural flavorings that were obtained from herbs that could be found in fields and woods or grown in one's garden. Pick one of the herbs below (or another that interests you), and write a report that includes some background on its native habitat, the mythology surrounding it, if any, the family to which it belongs, its medicinal use and its flavor. How is the herb used in cooking? Find out what the chemical structure of the flavoring agent is. The popular sources in Appendix A will help you get started.

   a. garlic (*Allium sativum*)

   b. peppermint (*Mentha piperita*), or another mint of the genus *Mentha*, (various species).

   c. sage (*Salvia officinalis*)

   d. rosemary (*Rosmarinus officinalis*)

   e. thyme (*Thymus vulgaris*)

INQUIRY 19

Acetylsalicylic Acid: For Your Aching Head

From ancient times humans have sought herbs which seemed to have magical power to relieve the pain of injury or of childbirth, to lower fever, to cure ailments, to stop bleeding, and to heal wounds. All so-called primitive cultures have had their special members, the medicine man or woman, the shaman, who knew just what herbs and perhaps what words would heal the patient's body (and sometimes the spirit as well). In our culture, the role of healing the body is filled by the medical doctor, and although in the twentieth century we like to think of medicine as pure science, it is a rare doctor who does not think of what he or she does as an art as well. Part of our science of medicine has come from the recognition in the nineteenth century that there were chemical substances in herbs that were responsible for their curative powers. One of those herbs was willow bark that contained a substance which, although it had irritating side effects, relieved pain and lowered fever—salicylic acid. In this Inquiry we will use commercial salicylic acid to synthesize a derivative of that acid, a familiar compound which does not have the side effects that salicylic acid had. Its name is acetylsalicylic acid—aspirin.

Equipment you will need

- balance
- beakers, various
- Büchner funnel
- Bunsen burner
- flask, Erlenmeyer, 125 mL
- ring stand and ring
- spatula
- suction flask and vacuum hose
- wire gauze

Other items

- filter paper
- paper towels

Chemicals you will use

- acetic anhydride, $C_4H_6O_3$
- ethanol, C_2H_5OH
- salicylic acid, $C_7H_6O_3$
- sulfuric acid, H_2SO_4, conc.
- water, distilled, H_2O

BEFORE YOU BEGIN

1. Define the following terms.

   a. aspirin:

b. analgesic:

c. antipyretic:

d. febrifuge:

e. anti-inflammatory:

2. Draw the structure of salicylic acid, $C_7H_6O_3$.

3. Draw the structure of aspirin, acetylsalicylic acid, $C_9H_8O_4$.

4. What is the relationship of acetic anhydride to acetic acid? Use their formulas to answer the question.

5. Write the equation for the formation of aspirin.

6. Your instructor will discuss the purpose of *recrystallization* in the synthesis of organic compounds such as aspirin. You may want to take notes.

SAFETY NOTES

1. Wear approved safety goggles at all times in the lab.

2. Acetic anhydride is IRRITATING to the skin and eyes. Dispense it in the hood. Maintain good ventilation in the lab. Concentrated sulfuric acid is CAUSTIC. If you get sulfuric acid or acetic anhydride on your skin or clothing, rinse immediately and tell your instructor. Do not remove the sulfuric acid dropper bottle from the reagent bench.

3. The aspirin you will make will probably be contaminated with some of the starting materials and, like all laboratory chemicals, MUST NOT BE TAKEN INTERNALLY.

4. All chemicals must be handled carefully and treated with respect. The solids and liquids used in this Inquiry are safe for you to use responsibly.

5. Wash your hands before you leave the lab.

PROCEDURE

1. Tear out the Inquiry 19 Observations page, and have it ready to take careful observations at each stage of the synthesis. Lay out a table for your weighings.

2. Set up a boiling water bath in a 600-mL beaker. (If you can use about 400 mL of distilled water, do so; the hot water can be used in step 6. If distilled water is limited, use tap water for the bath, and put about 25 mL of distilled water in a 100-mL beaker and warm it for step 6.)

3. In a 125-mL Erlenmeyer flask, weigh out about 1.25 g of salicylic acid, $C_7H_6O_3$, to the nearest 0.001 g. Add 2 mL of acetic anhydride, and 3 drops of concentrated sulfuric acid, H_2SO_4. Swirl the flask to mix the contents, and set the flask in the boiling water bath until the solid salicylic acid dissolves and reacts (10 minutes or more).

4. While the reaction is occurring, set up an ice bath in a 600-mL beaker, and get the Büchner funnel and suction flask ready for filtration.

5. Turn off the burner, carefully remove the hot flask, and set it in a beaker of ice water. You should see crystals of aspirin forming. When there is no more change, filter the crystals. The filtrate should be neutralized and discarded (see Cleanup).

6. To recrystallize the aspirin, remove the crystals from the funnel, and place them in a 100-mL beaker. Add 3-4 mL of ethanol to dissolve the crystals (you may have to set the flask containing the crystals back in the hot water bath for a few minutes). As soon as solution is complete, add about 15 mL of warm distilled water. Again, cool the crystals in the ice bath.

7. Refilter on a fresh filter paper, and air-dry for a few minutes under vacuum. Discard the filtrate, and remove the crystals and

place them on a clean paper towel. Spread them out, and press more water out of them with another towel.

8. Place the crystals in a 50-mL beaker. Your instructor may want you to leave the crystals until they are completely dry or may have you weigh them now.

CLEANUP

(All Cleanup suggestions are subject to local laws governing waste from laboratories. The following are suggestions only and may be changed by your instructor. Space has been provided for additional instructions.)

1. The aspirin you have produced may be put in the trash. All solutions should be checked with pH paper to be certain they are approximately neutral. If not, neutralize the waste acid with sodium carbonate or sodium bicarbonate and discard down the drain with running water.

2. Return all chemicals to their proper places in the lab, and wash and dry your lab bench. Wash your hands before you leave.

3. Additional instructions:

Name ________________________________ Sec ____________

INQUIRY 19: OBSERVATIONS

Name ______________________________ Sec__________

INQUIRY 19: RESULTS

1. What mass of aspirin did you produce?

2. In the reaction of salicylic acid with acetic anhydride to produce aspirin and acetic acid, the mole ratio of salicylic acid to aspirin is 1:1. Using your starting mass of salicylic acid, determine the theoretical yield of aspirin.

3. Compare your yield of aspirin with the theoretical yield by calculating the % yield:

$$\%\ \text{yield} = \frac{\text{actual yield}}{\text{theoretical yield}} \times 100$$

4. In the equation for the formation of aspirin, sulfuric acid is not a reactant. What was the purpose of the sulfuric acid? Where was it at the end of the reaction?

Name ______________________________ Sec ____________

INQUIRY 19: FOLLOW-UP QUESTIONS

These are to be done in the laboratory after the Inquiry. You are encouraged to discuss these with your lab partner and your lab instructor.

1. How does aspirin work to relieve pain? Refer to your textbook or other source.

2. What is a buffer? Does a "buffered" aspirin conform to the definition of "buffer"? Explain.

Special Projects

Your instructor may assign one or more of these Follow-up Special Projects. Some sources are given in Appendix A.

1. Look up the willow tree (family Salicaceae, genus *Salix*, various species) in a book on herbal medicine, and describe the medicinal properties of its bark. What is the chemical responsible for this?

2. Write a report on the meadowsweet (family Rosaceae, *Filipendula ulmaria* or *Spiraea ulmaria*). Describe the medicinal properties of this plant. What do you suppose is the source of the name "aspirin"? What chemical is the active principle in the meadowsweet?

3. Many modern medicines were originally extracted from plants. For example, digitalis, a cardiac stimulant used to treat congestive heart failure, was first extracted from foxglove, *Digitalis purpurea*, a plant which is cultivated in the United States as an ornamental. (Digitalis is also highly toxic and can cause a fatal heart attack! Poisonous herbs must be handled only by those trained to do so.) A major concern has been articulated by environmentalists and by some medical researchers that wanton destruction of our forests, especially the largely unexplored rain forests, may cause the extinction of some species of plants that could have been helpful to humankind. Find some articles on this subject, and write a short research paper.

INQUIRY 20

Soaps, Detergents, and Oil Spills

There is evidence that soap has been part of the human experience since antiquity. Some families today still have stories about great-grandma's soap kettle and the cleaning (and skin-drying!) power of her lye soap. In the 1960s and early 1970s the Hippie movement propelled many young people out of the cities to the countryside where they sought a more "authentic" experience of life away from what they perceived as a shallow, high-tech world of power-brokering. They became "back-to-the-landers," and, in communes or small communities, they began living deliberately, growing their own food, growing herbs for medicines, flavorings, and dyes (see Inquiries 18, 19, and 24), and making their own soaps and candles. Their interest helped renew a national interest in handcrafted items and a concern for keeping alive our cultural heritage. Beautiful handcrafted soaps, perfumed with herbs and colored with natural dyes, are today available in specialty shops.

As idyllic as life sometimes seemed to the back-to-the-landers, one thing many of them learned was that it is almost impossible to live solely on what one can produce oneself. Since some gave up their automobiles on becoming farmers and used horses for plowing and travel, the necessity for a job to bring in a little cash might mean a 20-mile trip on horseback or, what was more likely, getting the old car out of the barn and filling the tank with gasoline! And if one were given to thinking about the consequences of one's actions, one found that gasoline meant oil, that oil meant global economy, that global economy meant global environmental problems, and that global environmental problems meant global environmental solutions. Thus some back-to-the-landers came back to the world and became active environmentalists, addressing problems with an understanding that eliminating an essential but polluting industry, such as the oil industry, was not the answer. The answer lay partly in legislation to force industry to change, but, more importantly, it lay in cooperation between industry and environmentalists to produce a better world for all of us where problems like oil spills become a thing of the past.

The best way to deal with an oil spill is to prevent it, of course, but in the real world spills do occur, and that brings us full circle to soaps and detergents, which, though they don't solve problems of dead animals and destroyed habitats, can help to clean up the environmental mess. Inquiry 20 provides you an opportunity to make a soap, to make a controlled oil spill, and to examine detergents and soaps in that context.

Equipment you will need

beakers, various
Büchner funnel
droppers
evaporating dish
graduated cylinders, 10 and 50 mL
Petri dish halves (3)
suction flask and hose
test tubes (3), 100 mm
watch glass to cover evaporating dish

Chemicals you will use

ethanol, C_2H_5OH (optional)
sodium chloride, NaCl, sat'd
sodium hydroxide, NaOH, 20% by wt
water, distilled, H_2O

Other items

detergent samples, including commercial liquid laundry and dish detergents, laboratory detergents; wetting agents
ice
kerosene, vegetable oil, or mineral (paraffin) oil
lard
matches
motor oil, used
paper towels

BEFORE YOU BEGIN

1. Use your textbook to define the following terms

   a. surfactant:

   b. detergent:

   c. soap:

   d. saponification:

   e. surface tension:

   f. micelle:

2. Write the chemical equation for a general saponification reaction.

SAFETY NOTES

1. Wear approved safety goggles at all times in the lab.

2. If your instructor tells you to do Procedure I,A in which ethanol, C_2H_5OH, is used, remember that ethanol is very FLAMMABLE. If the ethanol catches fire, it will be confined to the beaker, and a wire gauze with a ceramic center can be placed over the beaker to extinguish the flame. Procedure I,B is provided for those who wish to avoid the hazard of using ethanol.

3. All chemicals must be handled carefully and treated with respect. The solids and liquids used in this Inquiry are safe for you to use responsibly.

4. Wash your hands before you leave the lab.

PROCEDURE

Part I: Making Soap

(Your instructor will assign you A or B.)

A. With ethanol

1. Tear out the Inquiry 20 Observations page, and have it ready to take observations at each step. Prepare about 100 mL of ice water for use in step 6.

2. Place a portion of lard about the size of 3 peas in a 100-mL beaker. Add 3 mL of ethanol, C_2H_5OH, and 3 mL of 20% sodium hydroxide, NaOH.

3. Begin heating gently while stirring. Using a small flame, hold the burner and sweep it around under the beaker, to heat gently and uniformly.

4. Heat gently until the ethanol has been driven off. (You should be able to detect the odor of ethanol by wafting the vapors toward your nose.) Then heat more vigorously, boiling the reaction mixture for about 5 minutes more. Cool the soap.

5. Add about 20 mL of saturated sodium chloride, NaCl, to "salt out" the soap. This will rinse away excess NaOH and glycerol formed by the reaction.

6. Filter using a Büchner funnel and vacuum filtration setup. Rinse the soap on the funnel using ice-cold water.

B. Without ethanol

1. Tear out the Inquiry 20 Observations page, and have it ready. Take organized observations throughout the Inquiry. Prepare about 100 mL of ice water for use in step 5.

2. Place a portion of lard about the size of 3 peas in an evaporating dish. Add 3 mL of 20% sodium hydroxide, NaOH.

Cover with a watch glass to prevent spattering. Begin heating gently.

3. Heat for 15 minutes, maintaining a low flame that sustains a gentle boil. When the soap is ready it will look white and uniform. Stop heating and cool the evaporating dish.

4. Pry the watch glass off carefully—it may be firmly attached to the evaporating dish with solidified soap. Scrape the soap out into a beaker containing 20 mL of saturated NaCl to "salt out" the soap. This will rinse away excess NaOH and glycerol formed by the reaction.

5. Filter using a Büchner funnel and vacuum filtration setup. Rinse the soap on the funnel using ice-cold water.

Part II: Testing Your Soap

1. Take a tiny piece of soap between your fingers and rub your fingers together under running water. How does it feel?

2. Put a tiny piece of your soap in a small test tube, and add distilled water. Cover the mouth of the test tube with your finger and shake vigorously. Observe.

3. Fill two 100-mm test tubes about 1/3 full of water. Add a few drops of kerosene, vegetable oil, or mineral (paraffin) oil to each test tube. Add a pinch of your soap to one of the tubes. Cover the mouth of each tube and shake it vigorously. Stand the tubes in a rack for a moment and observe.

Part III: Dealing With an Oil Spill

1. Obtain three Petri dish halves and place them on the black surface of your bench. Fill each with water.

2. Dip the wood end of a matchstick into used, dirty motor oil. Drain the excess oil off by touching the oil-coated end of the matchstick to a piece of paper towel.

3. Look at the water in one of the Petri dishes at an angle such that light is reflected off the surface of the water. Break the surface of the water with the oil-coated matchstick, then pull the matchstick out of the water. The result should be a slick that spreads out over the entire surface and looks like oil on a mud puddle. A spectrum should be visible. If the result was just a little puddle of oil in one spot, clean the dish with detergent, and refill it with water. The technique for producing this very thin layer of oil may require a little practice. One of the keys to this is having a *very* small amount of oil on the matchstick. When you have the technique perfected, put an oil slick on the other two water surfaces.

4. Fill three small test tubes about 1/3 full of water. Add a small piece of your soap to one, a drop of laundry or dish detergent to the second, and a drop of laboratory glassware detergent to the third. Shake each to mix well.

5. As you look at the one of the oil slicks, use a clean dropper to place one drop of the soap solution in the middle of the slick. What happens?

6. Repeat with the other two oil slicks, using a clean dropper to add a drop of prepared detergent to each of the slicks. Observe.

CLEANUP

(All Cleanup suggestions are subject to local laws governing waste from laboratories. The following are suggestions only and may be changed by your instructor. Space has been provided for additional instructions.)

1. The materials you have used are not harmful to the environment in small quantities and may be disposed of by rinsing them down the drain.

2. Return all chemicals to their proper places in the lab, and wash and dry your lab bench. Wash your hands before you leave.

3. Additional instructions:

Name ________________________________ Sec____________

INQUIRY 20: OBSERVATIONS

Name ______________________________________ Sec______________

INQUIRY 20: RESULTS

Draw conclusions from your observations. Evaluate your soap and its effectiveness in washing your hands and in altering the oil spill. Compare your soap to the detergents you tested.

Name ______________________________ Sec__________

INQUIRY 20: FOLLOW-UP QUESTIONS

These are to be done in the laboratory after the Inquiry. You are encouraged to discuss these with your lab partner and your lab instructor.

1. Use the general molecular structure of soaps to explain how a soap micelle forms.

2. One of the properties of bases such as NaOH is that they feel slippery, much like a soap feels on your fingers. Based on this Inquiry, suggest an explanation for this observation.

3. Use molecular structures to explain the difference between a soap and a detergent.

Special Projects

Your instructor may assign one or more of these Follow-up Special Projects.

1. Do a short research paper on detergents, concentrating on the ingredients of a detergent, types of surfactants, and biodegradability.

2. Write a short paper on the method used to clean beaches and shorelines after the *Exxon Valdez* oil spill in 1989. What are the long-term effects of such a spill?

3. Phosphates in detergents can cause water pollution. Write a short report to answer the following questions. What are phosphates? What purpose do phosphates serve in detergents? What is *eutrophication*? What compounds are being substituted for phosphates? Are they effective?

4. Interview an elderly person who remembers making his/her own soap (or perhaps who still does) or someone who makes soap and sells it at crafts fairs or shops. Write down the procedure used, noting the perfumes and colors added.

INQUIRY 21

Polymers: Those Big Molecules

If one word were to characterize innovation in materials in the last half of the twentieth century, it would probably be "plastics." No material has so completely revolutionized the way we live than these synthetic materials. The plastics that surround us, from fibers in clothes we wear, to seats in our cars, to paint on the walls of our houses, to the computer on which this paragraph was composed, to toothbrushes, telephones, teflon-coated pans—the list is limitless—all of these belong to a class of compounds called *polymers*. Polymers are very large molecules made from small molecules called *monomers*. In our age the word "plastic" has often been used to mean "polymeric material," but "plastic" really refers to the property of certain materials to be moldable into any shape. The synthetic plastics mentioned above are all polymers, but all polymers are not plastics. In this Inquiry you will take apart a familiar substance, starch, which is a natural polymer and is not plastic, and you will make some familiar substances which are synthetic polymers (and plastic).

Equipment you will need

beakers, various
Bunsen burner
copper wire (for hook)
disposable pipet
evaporating dish
graduated cylinder, 10 mL
ring stand and ring
stirring rod
test tubes, various
vial
wire gauze

Other items

bags, resealable, for slime
food coloring, various
paper towels
saliva
small piece of potato

Chemicals you will use

adipoyl chloride (4% by vol in hexane)
borax solution, 4% by wt
ethanol, C_2H_5OH
hexamethylenediamine (1,6-diaminohexane), (5% by wt in $0.1M$ NaOH)
iodine solution
poly(vinyl alcohol) soln, 4% by wt
starch, soluble

BEFORE YOU BEGIN

1. Define the following terms.

   a. starch:

   b. cellulose:

c. polymer:

d. monomer:

e. polysaccharide:

f. protein:

g. interface:

h. enzyme:

2. Look up in your text and describe the following processes:

   a. hydrolysis of starch (what are the products?):

   b. iodine test for starch (what would you see?):

SAFETY NOTES

1. Wear approved safety goggles at all times in the lab.

2. Concentrated hydrochloric acid, HCl, is very CAUSTIC. Adipoyl chloride is IRRITATING and must be handled carefully. If you get either of these chemicals on skin or clothing, rinse IMMEDIATELY with running water and inform your instructor.

3. While the HIV virus has been detected in all body fluids, there is no known case of transmission of the virus through saliva. However, to be safe from this and other, more easily transmitted viruses (such as hepatitis), it is a good idea for one partner to use his or her own saliva and to do all phases of the saliva hydrolysis (Procedure *and* Cleanup).

4. All chemicals must be handled carefully and treated with respect. The solids and liquids used in this Inquiry are safe for you to use responsibly.

5. Wash your hands before you leave the lab.

PROCEDURE

Tear out the Inquiry 21 Observations page, and have it ready to take organized observations at each step of the Procedure.

Part I: Preparation of Starch Suspension

1. Weigh out about 0.5 g of soluble starch in a 50-mL beaker, and add a few milliliters of water. Stir to make a paste. Add a few milliliters more of water, stir, and finally add water to the 40-mL mark. Set the beaker on a ring stand and heat to boiling. When the starch boils, turn off the burner.

2. Obtain about 10 mL of iodine solution in a vial, and put a dropper in it.

3. When the starch has cooled so that you can pour it from the beaker, pour a few milliliters into a 100-mm test tube. Add a few drops of iodine solution. Observe.

4. Obtain a piece of potato. Put a drop of the iodine solution on the potato. What happens?

Part II: Hydrolysis of Starch by Reaction with HCl

1. Set up a boiling water bath in a 100-mL beaker.

2. Since the next step requires that samples of the hydrolyzing starch be taken each minute, obtain a plastic, disposable pipet. Set up your Observations page for timing the reaction.

3. Put about 5 mL of starch suspension into a 100-mm test tube. Add 1 mL (or about 20 drops) of concentrated HCl. When you are ready to start timing the reaction and to take a sample each minute, place the test tube in the boiling water bath.

4. To test a sample each minute, withdraw a few drops of the hydrolyzing starch from the hot bath, and put it into a clean 100-mm test tube. Add 2 drops of iodine solution. Record your observations. Continue testing each minute until there is a change.

Part III: Hydrolysis of Starch by an Enzyme

1. Collect about 1 mL of your own saliva in a 150-mm test tube. Add about 10 mL of H_2O, and shake well.

2. Prepare to time the reaction. Pour 3 or 4 mL of the starch suspension into a 100-mm test tube. Add an equal volume of the saliva solution. Shake to mix. This is "time zero."

3. At the end of a minute, use a disposable pipet to withdraw a few drops of the starch-saliva mixture and place them in a 100 mm test tube. Add a drop or two of the iodine solution. Observe, and repeat every minute. Wash the test tubes and pipet which contained saliva before going on.

Part IV: Synthesis of Some Polymers

A. Nylon (interfacial polymerization)

1. Using a dropper, dispense about 20 drops of hexamethylenediamine solution into an evaporating dish. *Very carefully*, put 20 drops of the hexane solution of adipoyl chloride *down the side of the dish* so that the less dense hexane solution sits on top of the aqueous solution of the diamine and does not mix with it. This produces an interface at which the two solutions meet and at which polymerization takes place.

2. With a hooked copper wire, reach into the interface and hook the nylon that has formed there. Slowly pull the interface product out of the beaker, but do not touch it with your fingers. What is happening at the interface now?

3. When you have pulled all the nylon out, wash the fiber with a little ethanol followed by running water. (You may touch it after you have washed it well.)

4. Check the tensile strength by pulling on the fiber. See Cleanup to clean the evaporating dish.

B. Slime (cross-linking)

1. Obtain about 100 mL of poly(vinyl alcohol) solution in a 250-mL beaker. Add a few drops of your favorite food coloring, and then add 10 mL of 4% borax solution. Mix well.

2. With your fingers, remove the slime from the beaker. Pull it apart slowly and then rapidly. What happens?

3. Slime is nontoxic. You may take it home with you, unless your instructor tells you otherwise. (Do not give it to very small children.)

CLEANUP

(All Cleanup suggestions are subject to local laws governing waste from laboratories. The following are suggestions only and may be changed by your instructor. Space has been provided for additional instructions.)

1. Use soapy water to wash the evaporating dish you used to make nylon, and rinse with running water.

2. Slime must not be put into the sink (it will stop up the drain). The other materials you have used are not harmful to the environment in small quantities and may be disposed of by rinsing them down the drain.

3. Return all chemicals to their proper places in the lab, and wash and dry your lab bench. Wash your hands before you leave.

4. Additional instructions:

Name ______________________________ Sec __________

INQUIRY 21: OBSERVATIONS

Name ______________________________________ Sec______________

INQUIRY 21: RESULTS

Summarize your observations in two clear, concise paragraphs. Include in your summary a comparison of the rates of starch hydrolysis by HCl and by the enzyme(s) of saliva.

Name ____ Sec ______________

INQUIRY 21: FOLLOW-UP QUESTIONS

These are to be done in the laboratory after the Inquiry. You are encouraged to discuss these with your lab partner and your lab instructor.

1. Imagine that you have suddenly found that you must give up all plastics. Place yourself, either physically or mentally, in your home. List below all the things you will have to remove. (Don't forget to take the paint off the walls!)

2. The nylon you made was Nylon 6,6. It was made from 1,6-diaminohexane, $H_2N(CH_2)_6NH_2$ and adipoyl chloride, $ClOC(CH_2)_4COCl$. Using your text and these starting materials, draw the structure of Nylon 6,6.

3. Read the article on slime in the *Journal of Chemical Education* (see Appendix A). Draw the structure of slime. What is the purpose of borax?

4. Write a short paragraph explaining to a child why she or he should chew food slowly and completely before swallowing.

Name ______________________________________Sec______________

INQUIRY 21: FOLLOW-UP QUESTIONS (cont'd)

Special Projects

Your instructor may assign one or more of these Follow-up Special Projects. Some sources are given in Appendix A.

1. Slime has some unusual physical properties which make it a "non-Newtonian" fluid. Write a report on these special fluids.

2. Go to a local recyling center. How are polymers classified for recycling? What polymers can be recycled locally? What problems exist in polymer recycling?

3. Nylon was the first condensation polymer. Its history is a fascinating chapter in the commercial development of polymers in the first half of the twentieth century. Write a report detailing some of the research and the people involved.

4. Write a report on the digestion of polysaccharides, using structural formulas to show how starch is digested by our bodies and to explain why we are not able to digest cellulose.

INQUIRY 22

A Simple Complex: Analog of Natural Complexes

As you saw in Unit 2, metals play an integral part in our lives in their many forms as structural materials and as metallic compounds. But metals are essential in our own physical structure, as well. In our bodies metal ions may be found as the central atom in *coordination compounds* such as *hemoglobin*. Hemoglobin is a very large, complicated molecule, which contains an iron(II) ion complexed to the nitrogens of a molecule called a *porphyrin* (to form *heme*) and to the nitrogen of a protein that makes up most of the mass of the hemoglobin molecule. Interestingly, the *chlorophyll* molecule is a porphyrin with a magnesium atom at the center. It closely resembles the heme molecule. In this Inquiry we will extract chlorophyll from one of its common plant sources. Because chlorophyll is a complicated molecule, it is not possible for us to synthesize it. However, since complexation is such an important part of life processes, an analog from the inorganic world provides a simple complex ion, $Cu(NH_3)_4^{2+}$ [tetraamminecopper(II) ion)],which we can crystallize as the coordination compound, $Cu(NH_3)_4SO_4 \cdot H_2O$.

Equipment you will need

balance
beakers, various
capillary tube
conical funnel, filter paper
flask, 50 mL, with stopper to fit
graduated cylinder, 10 mL
heat lamp
mortar and pestle

Other items

paper towels
spinach leaves

Chemicals you will use

acetic acid, 3*M*
acetone, $(CH_3)_2CO$
ammonia, $NH_3(aq)$, conc, in dropper bottle
copper(II) sulfate pentahydrate, $CuSO_4 \cdot 5H_2O$
ethanol, C_2H_5OH, 95%
hydrochloric acid, 0.1*M*, standardized (optional)
hexane, C_6H_{14}
water, distilled, H_2O
sand, SiO_2

BEFORE YOU BEGIN

1. If you did Inquiry 11, you answered some questions about complex ions. Look at the Follow-Up Questions for Inquiry 11 (p. 91). If you did not answer question 2, do so here.

2. Use a general chemistry textbook (available in your library or from your instructor) and look up complex ions. Draw the structure of the tetraamminecopper(II) ion.

3. If you did not do Inquiry 18 your instructor will demonstrate how to set up paper chromatography. Take notes here.

SAFETY NOTES

1. Wear approved safety goggles at all times in the lab.

2. Concentrated ammonia, $NH_3(aq)$, is CAUSTIC and very IRRITATING to the mucous membranes of your nose and throat. Do not breathe it.

3. All chemicals must be handled carefully and treated with respect. The solids and liquids used in this Inquiry are safe for you to use responsibly.

4. Hexane and acetone are FLAMMABLE. Keep them away from flames.

5. Wash your hands before you leave the lab.

PROCEDURE

Part I: Tetraamminecopper(II) Sulfate

1. Tear out the Inquiry 22 Observations page, and have it ready to take organized observations.

2. Weigh out about 1 g of copper(II) sulfate pentahydrate, $CuSO_4 \cdot 5H_2O$, in a 50-mL beaker, and add 4-5 mL of distilled H_2O. Stir to dissolve, and, if necessary, heat gently.

3. If you have heated the solution, allow it to cool, and then, in the hood, add 3 droppersful of concentrated ammonia, $NH_3(aq)$. If a light blue precipitate results, add additional droppersful of ammonia solution until the light blue precipitate dissolves and the solution turns dark blue.

4. Add 5 mL of ethanol, C_2H_5OH. What happens? If there is no change, add a few additional milliliters of ethanol.

5. Set up the funnel for filtering, and pour your solution and crystals onto the filter paper. Rinse the beaker with a little

ethanol. Pour the rinsings onto the filter paper. Rinse again, if necessary, to obtain all the crystals.

6. Carefully take the filter paper out of the funnel, and open it up on a paper towel. Pat the crystals dry with the paper towel, or warm them under a heat lamp to evaporate the alcohol. Weigh the product. See Cleanup for disposal of the solid product and the ethanol in the filtrate.

Part II: Extraction of Chlorophyll from Spinach

1. Tear up two or three spinach leaves, and place them in a mortar. Add a small amount of sand to aid in grinding.

2. Place 4 mL of acetone, $(CH_3)_2CO$, in a 10-mL graduated cylinder, and add 1 mL of water. Pour this solvent on the spinach, and grind the spinach leaves with the pestle until the liquid has turned dark green.

3. Set up a conical funnel with a folded filter paper. Wet the filter paper with acetone. Decant the extracted material into the filter.

4. Using a capillary tube, spot the chromatography paper as demonstrated by your instructor. To concentrate the extract, apply at least 4 times, allowing the spot to dry after each application.

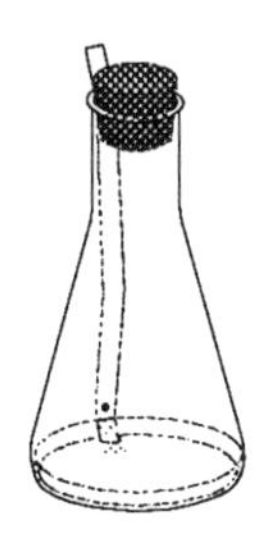

5. Place the chromatography paper in a 50-mL flask containing a few milliliters of hexane, C_6H_{14}, so that the spot of extract is just above the hexane. Stopper to hold the paper in place. Observe while the chromatogram develops. What is occurring?

6. To dispose of hexane and acetone, see Cleanup.

Part III: Optional

According to the formula for the copper coordination compound, there are 4 moles of ammonia molecules per mole of copper ions. To check to see if you obtained complete complexation you can titrate the tetraamminecopper(II) sulfate with standardized 0.1*M* hydrochloric acid. Note that the compound also contains one water molecule.

Write a procedure for determining the number of moles of ammonia per mole of compound. You can do either a microtitration like that in Inquiry 15, or you can use burets as in Inquiry 14. You will need to determine which indicator to use to titrate a weak base. Look in a chemical handbook for help. Your instructor may allow you to test some indicators to see which would be appropriate. In your procedure you should specify reagents and equipment you will need. Check your procedure with your instructor before you begin.

CLEANUP

(All Cleanup suggestions are subject to local laws governing waste from laboratories. The following are suggestions only and may be changed by your instructor. Space has been provided for additional instructions.)

1. Some chemistry departments may require recycling of ethanol. If not, the ethanol you used can go down the drain with water. Check with your instructor.

2. The amounts of hexane and acetone are so small that they may have evaporated before you clean up. If not, pour them into a beaker and set them in the hood to evaporate.

3. In the hood, dissolve the solid product in water and carefully add dilute acetic acid until the solution turns light blue. This will destroy the complex and neutralize the NH_3. Flush the solution down the drain with water.

4. Scrape the remainder of the spinach and sand into the trash.

5. The rest of the materials you have used are not harmful to the environment in small quantities and may be disposed of by rinsing them down the drain.

6. Return all chemicals to their proper places in the lab, and wash and dry your lab bench. Wash your hands before you leave.

7. Additional instructions:

Name ______________________________ Sec__________

INQUIRY 22: OBSERVATIONS

Take all observations in an organized fashion. Make tables for weighings, and attach your chlorophyll chromatogram to this page.

Name ______________________________ Sec__________

INQUIRY 22: RESULTS

Part I

1. $CuSO_4 \cdot 5H_2O$ reacted with NH_3 to produce $Cu(NH_3)_4SO_4 \cdot H_2O$. Write the equation for the reaction.

2. From the equation, calculate the theoretical yield of the coordination compound.

3. Calculate your % yield:

$$\% \text{ yield} = \frac{\text{actual yield}}{\text{theoretical yield}} \times 100$$

Part II

1. a. The yellow dye on your chromatogram is a mixture of carotenoids. What is a carotenoid?

   b. Name a common carotenoid.

2. Suggest a reason why the carotenoids and chlorophylls separated as they did with the nonpolar solvent, hexane.

3. Refer to Appendix D, and name another common carotenoid.

Part III: Optional

From the titration data, calculate the number of moles of ammonia, NH_3, in one mole of the coordination compound you synthesized. How does your answer compare to the formula?

Name ______________________________ Sec__________

INQUIRY 22: FOLLOW-UP QUESTIONS

These are to be done in the laboratory after the Inquiry. You are encouraged to discuss these with your lab partner and your lab instructor.

1. Find the chemical structure of chlorophyll in a general chemistry or biochemistry textbook, and draw it here. Is there more than one structure for chlorophyll?

2. How are the two coordination compounds in this Inquiry [tetraamminecopper(II) sulfate and chlorophyll] similar? Look specifically at the way each coordinates to nitrogen atoms.

3. a. How are the carotenoids related to vitamins?

   b. Which vitamin has a carotenoid as a precursor in some vegetables?

   c. What is "night blindness," and how is it related to the carotenoids?

Name ______________________________________ Sec______________

INQUIRY 22: FOLLOW-UP QUESTIONS (cont'd)

Special Projects

Your instructor may assign one or more of these Follow-up Special Projects. Some sources are given in Appendix A.

1. Write a report on hemoglobin. How is iron bonded to the prophyrin molecule? How is oxygen carried by the iron? What happens chemically when a person breathes carbon monoxide?

2. Write a report on the chemical reactions in photosynthesis, concentrating on the role of chlorophyll in this process.

Unit 5

Chemistry for the Future

It is impossible for us to predict what the future will bring, but we can be certain that chemistry will have a part in it. Whether we will have the resources available to us that the crew on the Starship *Enterprise* has remains to be seen. It's exciting to imagine a universe in which we can travel at warp speeds; in which we can dematerialize the human body, beam it as pure energy to another location, and rematerialize it with all the molecules in the right places; and where, if we have aches and pains after energy travel, we can request an aspirin and a glass of water, and the replicator in our quarters will materialize these items from their molecular codes. What a limitless future may await us!

This last Unit, however, is not about chemistry *of* the future, but about using chemistry *for* the future, to preserve the Earth for ourselves and our descendants. We are daily made aware that it is irresponsible to continue being a throwaway society, that recyling our resources is essential. Many of those resources are metals, and in the last Inquiries you will address questions of metal ion effluents from industry and metal recycling.

INQUIRY 23a

A Problem: To Analyze an Industry Effluent (Ag^+)

Your consulting firm has been contracted by Fancy Silverplating, Inc., to help them solve a vexing problem. They have discovered that they are losing silver solution at an alarming rate. They believe that it is leaking into the sanitary sewage system from some very old piping. Not only is this an expensive problem, but the Environmental Protection Agency (EPA) has become interested, because the concentration of Ag^+ ion in the effluent is exceeding standard limits. Fancy's President, Mary F. Ware, is concerned about stiff fines which Fancy can't afford.

You will be given a sample of effluent which you will analyze by gravimetric means, precipitating silver chloride, and, applying your knowledge of chemistry, determining by mole ratio the amount of silver ion in the original sample. Then you must report your findings to President Ware on a report page of your own design. Your report should be accompanied by a cover letter to President Ware, summarizing your findings.

Equipment you will need

balance
beakers, various
Bunsen burner
conical funnels (3)
filter paper
oven (optional)
pipet, volumetric, 10 mL
pipet pump or bulb
ring stand and ring
stirring rod
test tubes (3) ,150 mm
wash bottle
watch glasses (3)
wire gauze

Chemicals you will use

effluent solution to analyze
nitric acid, HNO_3, 6 *M*
sodium chloride, NaCl, 1 *M*
water, distilled, H_2O

Other items

boiling chips or spheres

BEFORE YOU BEGIN

1. Define *effluent.*

2. What is meant by the abbreviation *ppm*?

3. Define *gravimetric*.

4. Your instructor will have some specific instructions about pipeting, filtering, washing the precipitate, and drying the precipitate. Take notes here.

SAFETY NOTES

1. Wear approved safety goggles at all times in the lab.

2. Silver compounds will stain skin and clothing.

3. All chemicals must be handled carefully and treated with respect. The solids and liquids used in this Inquiry are safe for you to use responsibly.

4. Wash your hands before you leave the lab.

PROCEDURE

1. Tear out the Inquiry 23 Observations page, and have it ready. This page is only for your rough notes; you must record observations carefully since your final report will come from your notes.

2. Obtain a sample of effluent from your instructor. This sample must last through three determinations, so take care not to waste it.

3. Heat a water bath in a 250-mL beaker. Put a boiling chip in the water. While the water is approaching the boiling point, go on to step 4. When the water boils you can turn the burner off.

4. Practice using the volumetric pipet by delivering 10 mL of distilled water to a beaker until you feel comfortable with the technique. When you are ready to pipet the effluent, rinse the pipet by drawing a few milliliters of effluent sample into the pipet and rolling the solution around in the pipet. Do this twice, disposing of the used solution in a specially designated waste beaker. Why is rinsing with the effluent solution necessary?

5. Pipet 10 mL of effluent into each of three clean 150-mm test tubes. To each test tube add 3 drops of $6M$ HNO_3 and 5 mL of $1M$ NaCl.

6. Stir each AgCl precipitate, taking care to use a very small amount of water from a wash bottle to rinse precipitate clinging to the stirring rod back into the test tube, before stirring the next precipitate. Set the test tubes in the water bath and maintain the temperature just under boiling for 5 minutes to help the precipitates coagulate.

7. Allow the precipitates to settle, and then put an additional drop of NaCl into one of the tubes to check for complete precipitation. If more AgCl forms, add 1 mL of NaCl solution to each test tube. If no precipitate forms, precipitation is complete.

8. Carefully decant the supernatant liquids in each test tube into a beaker. Wash each precipitate with distilled water and a drop of HNO_3, stirring to break up the clumps.

9. Wash the precipitates a total of 4 times, adding the drop of HNO_3 each time. The final time, set up three funnels with *weighed* filter papers, and prepare three watch glasses with a slip of paper identifying the sample number. Filter each precipitate, washing remaining solid out of each test tube with a stream of water from the wash bottle.

10. Place the watch glasses containing the filter papers with precipitates into an oven at 110°C for an hour, or in your locker until next lab period. When the precipitates are dry, weigh them. Calculate the molarity of the original sample, and report your results as ppm Ag^+ in the effluent.

CLEANUP

(All Cleanup suggestions are subject to local laws governing waste from laboratories. The following are suggestions only and may be changed by your instructor. A space has been provided for additional instructions.)

1. As a consultant dealing with environmental concerns you would want to be part of the solution and not part of the problem, so dispose of your silver waste by the most economical method of disposal, which is, of course, recycling. Place your AgCl precipitates and any waste effluent in the appropriate container for use in Inquiry 25.

2. The supernatant liquids you decanted are acidic. Place the beaker containing the waste solution in the sink and slowly add either Na_2CO_3 or $NaHCO_3$ solution until the "fizzing" stops. Pour the resulting solution down the drain with running water.

3. The other materials you have used are not harmful to the environment in small quantities and may be disposed of by rinsing them down the drain.

4. Return all chemicals to their proper places in the lab, and wash and dry your lab bench. Wash your hands before you leave.

5. Additional instructions:

Name ____Sec______________

INQUIRY 23a: OBSERVATIONS

Name ____Sec_______________

INQUIRY 23a: RESULTS

Your report should be typed and look professional. You may design letterhead for your company, if you like.

INQUIRY 23b

A Problem: To Analyze an Industry Effluent (Cu^{2+})

Your consulting firm has been contracted by Shiny Copperplating, Inc., to help them solve a vexing problem. They have discovered that they are losing copper solution at an alarming rate. They believe that it is leaking into the sanitary sewage system from some very old piping. Not only is this an expensive problem, but the Environmental Protection Agency (EPA) has become interested, because the concentration of Cu^{2+} ion in the effluent is exceeding standard limits. Shiny's President, William C. Penny, is concerned about stiff fines which Shiny can't afford.

You will be given a sample of effluent which you will analyze by gravimetric means, precipitating copper (II) oxide and, applying your knowledge of chemistry, determining by mole ratio the amount of copper ion in the original sample. Then you must report your findings to President Penny on a report page of your own design. Your report should be accompanied by a cover letter to President Penny, summarizing your findings.

Equipment you will need

balance
beakers, various
Bunsen burner
conical funnels (3)
filter paper
oven (optional)
pipet, volumetric, 10 mL
pipet pump or bulb
ring stand and ring
stirring rod
test tubes (3), 150 mm
wash bottle
watch glasses (3)
wire gauze

Chemicals you will use

effluent solution to analyze
sodium hydroxide, NaOH, 6*M*
water, distilled, H_2O

Other items

boiling chips or spheres
pH paper

BEFORE YOU BEGIN

1. Define *effluent.*

2. What is meant by the abbreviation *ppm*?

3. Define *gravimetric*.

4. Your instructor will have some specific instructions about pipeting, filtering, washing the precipitate, and drying the precipitate. Take notes here.

SAFETY NOTES

1. Wear approved safety goggles at all times in the lab.

2. All chemicals must be handled carefully and treated with respect. The solids and liquids used in this INQUIRY are safe for you to use responsibly.

3. Wash your hands before you leave the lab.

PROCEDURE

1. Tear out the Inquiry 23 Observations page and have it ready. This page is only for your rough notes; you must record observations carefully, since your final report will come from your notes.

2. Obtain a sample of effluent from your instructor. This sample must last through three determinations, so take care not to waste it.

3. Heat a water bath in a 250-mL beaker. Put a boiling chip in the water. While the water is approaching the boiling point, go on to step 4. When the water boils you can turn the burner off.

4. Practice using the volumetric pipet by delivering 10 mL of distilled water to a beaker until you feel comfortable with the technique. When you are ready to pipet the effluent, rinse the pipet by drawing a few milliliters of effluent sample into the pipet and rolling the solution around in the pipet. Do this twice, disposing of the used solution in a specially designated waste beaker. Why is rinsing with the effluent solution necessary?

5. Pipet 10 mL of effluent into each of three clean 150 mm-test tubes. To each test tube add 3 mL of 6 *M* NaOH.

6. Stir each $Cu(OH)_2$ precipitate, taking care to use a very small amount of water from a wash bottle to rinse precipitate clinging to the stirring rod back into the test tube, before stirring the next precipitate. Set the test tubes in the water bath and maintain the temperature just under boiling for 5 minutes to help the precipitates coagulate. What happens as the temperature increases? What is the formula of the new species?

7. Allow the precipitates to settle, and then put an additional drop of NaOH into one of the tubes to check for complete precipitation. If more solid forms, add 1 mL of NaOH solution to each test tube. If no precipitate forms, precipitation is complete.

8. Set up three funnels with *weighed* filter papers, and prepare three watch glasses with a slip of paper identifying the sample number. Filter each precipitate, washing remaining solid out of each test tube with a stream of water from the wash bottle.

9. Rinse the precipitates with a stream of water from the wash bottle. After they have drained, rinse again. Continue to rinse and drain until a test of the filtrate from one of the funnels is neutral to pH paper.

10. Place the watch glasses containing the filter papers with precipitates into an oven at 110°C for an hour, or in your locker until next lab period. When the precipitates are dry, weigh them. Calculate the molarity of the original sample, and report your results as ppm Cu^{2+} in the effluent.

CLEANUP

(All Cleanup suggestions are subject to local laws governing waste from laboratories. The following are suggestions only and may be changed by your instructor. Space has been provided for additional instructions.)

1. As a consultant dealing with environmental concerns you would want to be part of the solution and not part of the problem, so dispose of your copper waste by the most economical method of disposal, which is, of course, recycling. Place your CuO precipitates and any waste effluent in the appropriate container for recycling.

2. The filtrate from the CuO is basic. Dilute it by flushing it down the drain with running water.

3. The other materials you have used are not harmful to the environment in small quantities and may be disposed of by rinsing them down the drain.

4. Return all chemicals to their proper places in the lab, and wash and dry your lab bench. Wash your hands before you leave.

5. Additional instructions:

Name ______________________________ Sec____________

INQUIRY 23b: OBSERVATIONS

Name ______________________________________Sec______________

INQUIRY 23b: RESULTS

Your report should be typed and look professional. You may design letterhead for your company, if you like.

INQUIRY 24

The Ubiquitous Aluminum Can: Dyeing to Recycle

Aluminum is the third most abundant element in the Earth's crust. Because of its light weight (its density is 2.7 g/cm^3), it is in demand as a structural metal in airplane construction and related industries. Aluminum is refined from its ore, bauxite, using its property of amphoterism (see Before You Begin) to purify the alumina (Al_2O_3) in the ore. Electrolysis is then used to produce pure aluminum. The recycling of aluminum has become a major effort, saving the environment from beer and soft drink cans that take hundreds of years to decompose and producing new aluminum from old for about 5% of the cost of refining aluminum ore. Aluminum compounds are found in toothpaste, deodorant, antacid tablets, and baking powder. They are used in water purification, in pickling vegetables—in all areas of our lives. One of those areas is in the colorful clothing we wear. A compound of aluminum, potassium aluminum sulfate, commonly called alum, can be used as a mordant in the dye process. In this Inquiry, we will produce alum from a discarded aluminum can and then use the alum with natural dyes from onion skin and other plants (and perhaps a synthetic dye or two) to tie-dye a piece of cloth or clothing.

Equipment you will need

balance
beakers, various
Büchner funnel
Bunsen burner
conical funnel
filter flask and vacuum hose
forceps
graduated cylinder, 10 mL
hot plate
oven (optional)
ring stand and ring
rubber policeman
tin snips
watch glass
wire gauze

Chemicals you will use

aluminum can from Inquiry 17
calcium oxide, CaO
ethanol, 50% by vol
potassium hydrogen tartrate, $KHC_4H_4O_6$ (cream of tartar)
potassium hydroxide, KOH, 2*M*
sodium hydroxide, NaOH, 0.1*M*
sulfuric acid, H_2SO_4, 6*M*

Optional dyes to try

(see Appendix D)

alizarin
turmeric

Other items

ice
pieces of 100% cotton cloth, 1 x 1 inch for tests, 1 x 1 foot for tie-dyeing
string
yellow onion skins (and other vegetables and herbs to try, such as red onion, red cabbage)

BEFORE YOU BEGIN

1. Define the following:

   a. bauxite

   b. amphoteric

   c. dye

   d. mordant

2. The equations for the formation of alum are as follows:

$$2Al(s) + 2K^+(aq) + 2OH^-(aq) + 6H_2O \rightarrow 2K^+(aq) + 2Al(OH)_4^-(aq) + 3H_2(g)$$

$$2K^+(aq) + 2Al(OH)_4^-(aq) + 4H_2SO_4(aq) \rightarrow 2K^+(aq) + 2Al^{3+}(aq) + 4SO_4^{2-}(aq) + 8H_2O$$

$$K^+(aq) + Al^{3+}(aq) + 2SO_4^{2+}(aq) + 12H_2O \rightarrow \underset{\text{(alum)}}{KAl(SO_4)_2 \cdot 12H_2O(s)}$$

SAFETY NOTES

1. Wear approved safety goggles at all times in the lab.

2. Potassium hydroxide, KOH, and sulfuric acid, H_2SO_4, are very CAUSTIC. The solutions you are using are fairly concentrated, so handle them with EXTREME CARE. If you get any of either solution on your skin or clothing, rinse with copious amounts of water IMMEDIATELY, and tell your instructor.

3. The reaction of Al with KOH produces hydrogen, $H_2(g)$, a very FLAMMABLE gas. The reaction must be heated on a hot plate in a hood.

4. All chemicals must be handled carefully and treated with respect. The other solids and liquids used in this Inquiry are safe for you to use responsibly.

5. Wash your hands before you leave the lab.

PROCEDURE

If the Inquiry is scheduled for two lab periods, be sure to bring an item to tie-dye. Items to be dyed should be white and 100% cotton, such as a handkerchief, a headband (i.e., any item which will fit in a large beaker!) Additional white cotton cloth will be available from your instructor.

Part I: Making Potassium Aluminum Sulfate (Alum)

1. Tear out the Inquiry 24 Observations page, and have it ready. Take observations, including weighings, in an organized fashion.

2. Cut very small pieces off the aluminum can, and weigh out about 0.2 g of aluminum to the nearest 0.001 g. Place the Al pieces in a 100-mL beaker. Add 10 mL of the KOH solution. There should be several hot plates available in the hood. Mark your beaker with your initials and place it in the hood on a hot plate. Heat very gently until the aluminum is gone. (While you are heating the mixture, obtain and set up the Büchner funnel for step 6.) There may be small amounts of non-Al material floating around. When the reaction has stopped, allow the solution to cool. (Cooling can be accelerated by setting the 100-mL beaker in a larger beaker of cold water.)

3. While the solution is cooling, set up a conical funnel with a piece of coarse filter paper. When the beaker is cool enough to touch, carefully pour the warm solution through the filter paper to trap the undissolved solids. (What are they?)

4. Slowly add 6 mL of 6 *M* H_2SO_4 to the solution with stirring. What is happening? Heat the mixture gently over the Bunsen burner for about 10 minutes. What happens? While the mixture is heating, set up the ice bath in step 5. Cool the beaker for a few minutes after heating.

5. Put about 100 mL of ice in a 600-mL beaker. Add water to the level of the ice. Place the beaker of solution from step 4 on the ice for about 15 minutes. While you are waiting for the alum crystals to form, set up a Büchner funnel with a *weighed* filter paper.

6. Transfer the crystals to the Büchner funnel, scraping the last crystals from the funnel with a rubber policeman. Wash with 5-mL portions of 50% ethanol solution several times to remove all soluble ions and to begin drying the alum. When the last rinse is done, let the crystals dry for several minutes while the vacuum is running to make them easier to remove from the funnel.

7. Break the vacuum, and carefully remove the filter paper and crystals together, and place them on a watch glass. Your instructor will tell you whether to put them in your locker to air dry for weighing next week or to put them in a warm (<80°C) oven for drying.

8. Weigh the crystals and filter paper when dry, and save the crystals for step 9 and Part II. Save the filtrate for Cleanup.

9. Dissolve a small amount (match head size) of your alum in water in a small test tube. Add 0.1*M* NaOH dropwise until there is no more change. Observe. What is this property called?

Part II: Testing a Dye

1. Put about 50 mL of water in a 100 mL beaker, and add a pea-sized amount of the alum you made and a smaller amount (match head) of cream of tartar, $KHC_4H_4O_6$. This is the mordant solution. Add one or two small (1 x 1 inch) pieces of cotton cloth and boil for about 2 min.

2. While the mordant solution in 1 is boiling, set another 100-mL beaker containing about 50 mL of water over a Bunsen burner, put onion skin from about 1/4 of an onion in the water, and boil for 5 minutes.

3. Wet a piece (1 x 1 inch) of *un*mordanted cloth and place it in the onion solution and boil for a minute. Using forceps, remove the cloth and rinse it well under running water. Observe.

4. When the mordanted cloth is ready, use forceps to lift a piece out of the solution and rinse it under running water. Place the mordanted cloth in the onion solution and boil for 1 minute. Remove it and rinse it well under running water. Observe. Dry the cloth samples, and attach them to your report with labels.

Part III: Tie-dye

Natural dyes require some experimentation, as the chemical makeup of an herb can depend on the soil in which it is grown. Some sources are given in Appendix A for more reading in this area. Check your local libraries for others. These instructions, using alizarin, will give you an introduction to tie-dyeing. (For more on alizarin, see Appendix D.)

1. Lay out the cloth, and make some ties according to the instructions in Appendix C.

2. Prepare a mordant bath in a 250-mL beaker of water, using about 2 peas worth of alum and the same amount of CaO. Add the cloth you have tied, and boil for 5 minutes.

3. While the cloth is mordanting, place about 0.2 g alizarin in a 250-mL beaker, and add about 100 mL H_2O. Boil for 2 minutes.

4. Place cloth in the dye solution and boil for 5 minutes. Turn off the burner, and take out the cloth and rinse well in cold water. Remove the ties and rinse again.

Optional

1. Repeat the dye tests and tie-dyeing with red cabbage, which works well with the alum mordanting process in Part II.

2. Try tie-dyeing with turmeric, a common spice. Set up a boiling water bath for turmeric as above. It does not require mordanting. See Appendix D for more information on turmeric.

CLEANUP

(All Cleanup suggestions are subject to local laws governing waste from laboratories. The following are suggestions only and may be changed by your instructor. Space has been provided for additional instructions.)

1. The materials you have used are not harmful to the environment in small quantities and may be disposed of by rinsing them down the drain.

2. Return all chemicals to their proper places in the lab, and wash and dry your lab bench. Wash your hands before you leave.

3. Additional instructions:

Name ______________________________ Sec____________

INQUIRY 24: OBSERVATIONS

Name ______________________________________Sec______________

INQUIRY 24: RESULTS

Attach your dye tests to this page, and draw conclusions about the use of alum as a mordant.

Name ______________________________ Sec __________

INQUIRY 24: FOLLOW-UP QUESTIONS

These are to be done in the laboratory after the Inquiry. You are encouraged to discuss these with your lab partner and your lab instructor.

1. How is alum used in water treatment plants? Look in your textbook or in an encyclopedia.

2. In a recycling station, what simple procedure is used to separate aluminum from steel cans?

3. The dye responsible for the yellow color of onion skins is called *quercetin*. It provides the yellow color of many other plants as well, including tea, sumac, and a species of oak. Quercetin's derivatives (other compounds formed on this basic structure) are found in the cotton flower, in maize, and in some golden apples.* Its chemical name is 3,3',4',5,7-pentahydroxyflavone.

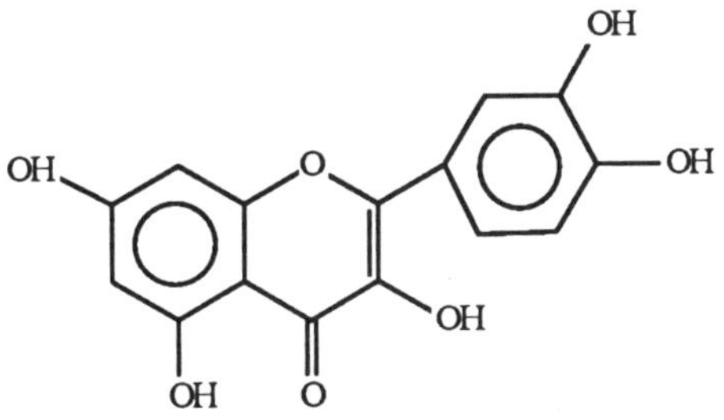

   Quercetin can be easily isolated as yellow crystals with a melting point of 316°C. You have already extracted quercetin from the onion skins. Assuming that there are no other substances in the solution you made, write a procedure for obtaining the pure crystals and detail one method of confirming its identity. (Refer to Inquiry 4 for help.)

4. The quercetin molecule contains three different functional groups. Name them.

---

* Mayer, Fritz. *Chemistry of Natural Coloring Matters*, American Chemical Society Monograph 89, translated and revised by A. H. Cook. New York: Reinhold, 1943.

Name ______________________________ Sec ____________

INQUIRY 24: FOLLOW-UP QUESTIONS (cont'd)

5. The aluminum ion reacts with water to form an acidic solution. Potassium hydrogen tartrate, cream of tartar, also reacts with water to produce an acid solution by the following reaction:

$$KHC_4H_4O_6(s) + H_2O \rightarrow K^+(aq) + H_3O^+(aq) + C_4H_4O_6^{2-}(aq)$$

When Al^{3+} complexes with the dye quercetin, it requires a particular pH. From the information and equation above, what pH range must be necessary?

6. Calcium oxide reacts with water to produce a base, $Ca(OH)_2$. What pH range must be necessary in the mordanting of alizarin?

7. Use your text or another source to answer the following questions.

   a. How does a dye molecule attach to a fiber?

   b. What is the purpose of the mordant?

Special Project

Your instructor may assign one or more of these Follow-up Special Projects.

1. Go to a local industry and interview management and labor. Write a report on ways in which the industry is recycling to save money and to prevent waste from entering the local water table.

2. Write a report on natural dyes, detailing the chemistry of several of the dyes. The resources mentioned may help you.

3. Using one of the resources mentioned in Appendix A, experiment at home with various natural dyes, and tie-dye some clothing. Keep careful notes of each experiment so that successful procedures could be used by others in your class. Turn in your tie-dye and your procedures for credit.

4. The *flavones* form an interesting group of dye compounds. Do a report on these dyes.

5. The one dye which is probably most prevalent in any college classroom is *indigo*. It is the blue of blue jeans. Write a report on indigo, its history, its structure, and its use in the denim industry. Where does indigo occur naturally? How is the dye synthesized?

INQUIRY 25

Reclaiming Silver

From the silver mines at Laurion, Greece, which provided ancient Athens with coinage silver, to the Comstock Lode in Nevada, silver has been one of a handful of metallic elements called the *precious metals*. Like gold, silver was prized by the ancients for coins and jewelry. Like gold, it spawned whole towns in the western United States in the nineteenth century. And, also like gold, it became harder to find as mines were worked out and nuggets became rarer. Silver is now mined principally as the mineral argentite, Ag_2S, which must undergo extensive chemical processing to produce silver. Though not valued as highly as gold, silver's price, like gold's, fluctuates in the open market. For this reason, the price of silver compounds fluctuates as well, especially when the silver market is very active. And because silver is relatively expensive, its compounds are expensive. For example, in 1990 dollars, 500 g (just over a pound) of copper(II) nitrate could be purchased for $34.10, while 500 g of silver nitrate of equal purity listed for $305.15.* Not only are silver compounds expensive, silver ion is toxic and should not be discarded in the sanitary sewage system. For these reasons silver ion is collected from industrial processes, such as photographic developing, and recycled. A twenty-first century procedure for producing silver, then, is less likely to be in the form of "staking a claim" for new silver than it is to be "reclaiming" silver from waste.

Equipment you will need

beakers, various
funnel, conical
oven or heat lamp
stirring rod
wash bottle
watch glass

Chemicals you will use

silver chloride from various Inquiries
sodium carbonate, Na_2CO_3, for waste treatment
sulfuric acid, H_2SO_4, 3 *M*
water, distilled, H_2O
zinc, Zn, granulated

Other items

filter paper

BEFORE YOU BEGIN

1. Although most of the silver waste was deposited in the waste containers as AgCl or $AgNO_3$, there may have been some small amounts of Ag_2CO_3, AgOH, and Ag_3PO_4 present. Your instructor has converted the silver waste which you deposited in the waste container this semester to AgCl by dissolving each of these solids with H_2SO_4, and then precipitating AgCl by adding excess NaCl.

* Aldrich Chemical Company, Inc., *Catalog Handbook of Fine Chemicals*. Milwaukee: 1990.

a. Write the equations for the reactions of Ag_2CO_3 and AgOH with H_2SO_4. (If you are unsure about the products, look up reactions of H_2SO_4 with any metal carbonate or hydroxide in the acid-base chapter of your text or in a previous Inquiry.)

b. Write the net ionic equation for the reaction of $AgNO_3$ and NaCl.

2. It is possible that the mass of silver chloride available to you will be more or less than that specified in the Procedure. If you are asked to change the amounts of reagents used, mark through the amounts given in the Procedure, noting new quantities in the margin.

SAFETY NOTES

1. Wear approved safety goggles at all times in the lab.

2. Silver compounds will cause a stain on skin and clothing. Always wash your hands immediately after working with silver compounds.

3. When sulfuric acid, H_2SO_4, is added to zinc, Zn (step 4 below) hydrogen is produced. THIS MUST BE DONE IN THE HOOD. NO FLAMES SHOULD BE PRESENT.

4. All chemicals must be handled carefully and treated with respect. The solids and liquids used in this Inquiry are safe for you to use responsibly.

5. Wash your hands before you leave the lab.

PROCEDURE

1. Tear out the Inquiry 25 Observations page, and be prepared to take observations as required.

2. Weigh out about 1 g of the solid waste silver chloride to the nearest 0.001 g, and transfer the powdered AgCl to a 50-mL beaker. Weigh out about 0.5 g of Zn and transfer it to the 50-mL beaker with the solid AgCl. Take the beaker to the hood.

3. WORKING IN THE HOOD, add 4 mL of $3M$ H_2SO_4. Stir until the Zn is *completely* used up. What is happening?

4. Decant the supernatant liquid into a 100-mL beaker which will act as a waste container. What is in the solution? What remains behind in the 50-mL beaker?

5. Add another 4 mL of $3M$ H_2SO_4, and stir for a minute or two. Decant again into the waste beaker.

6. Set up a conical funnel with a *weighed* folded filter paper. Add 5 mL of distilled water to the solid in the 50-mL beaker. Describe the solid. Stir to rinse, and decant this time into the funnel, holding the solid in the beaker and collecting the filtrate in the waste beaker. Again add 5 mL of distilled water to the solid, and decant once more into the funnel, retaining the solid in the beaker. Finally, add a last 5-mL portion of distilled water, and this time pour liquid and solid onto the filter paper. Rinse any remaining silver into the funnel, and rinse your stirring rod over the funnel with a small stream of water to remove the last traces of silver. After the solution runs through the funnel, rinse the solid on the paper with a stream of water from a wash bottle.

7. Carefully lift the filter paper out of the funnel and place it on a watch glass. Dry the filter paper and Ag in an oven (110°C) or under a heat lamp. When the Ag is dry, weigh paper + Ag, note the mass, and scrape the silver off the paper into a bottle designated by your instructor. Put the paper in the trash.

CLEANUP

(All Cleanup suggestions are subject to local laws governing waste from laboratories. The following are suggestions only and may be changed by your instructor. A space has been provided for additional instructions.)

1. The waste solution contains low toxicity zinc salts and excess sulfuric acid. Neutralize it slowly in the sink with some Na_2CO_3 or $NaHCO_3$ and pour the neutralized solution down the drain with running water.

2. Return all chemicals to their proper places in the lab, and wash and dry your lab bench. Wash your hands before you leave.

3. Additional instructions:

Name ______________________________ Sec__________

INQUIRY 25: OBSERVATIONS

Name ____Sec______________

INQUIRY 25: RESULTS

1. Calculate the mass of the Ag you produced.

2. The reaction of AgCl with Zn is

$$AgCl(s) + Zn(s) \rightarrow Ag(s) + Zn^{2+}(aq) + 2Cl^{-}(aq)$$

   Calculate the theoretical yield of Ag from the initial mass of AgCl.

3. How did your actual yield compare with the theoretical yield? Account for any difference.

Name ______________________________________ Sec____________

INQUIRY 25: FOLLOW-UP QUESTIONS

These are to be done in the laboratory after the Inquiry. You are encouraged to discuss these with your lab partner and your lab instructor.

The chemistry storeroom can use the silver you made to produce $AgNO_3$ for next semester's Inquiries. The reaction is:

$$Ag(s) + 2HNO_3 \rightarrow AgNO_3 + 2NO_2(g) + H_2O$$

1. From the mass of Ag you made, how much $AgNO_3$ can the storeroom make next semester?

2. At the 1990 price of $AgNO_3$ quoted in the introduction to this Inquiry, how much money did you save the storeroom? (Assume that the cost of HNO_3 is negligible compared to the cost of Ag compounds.)

3. a. Use a dictionary to find the definition of *troy weight.* How many troy ounces are there in a troy pound?

   b. What is *avoirdupois (avdp.) weight*? How many avdp. ounces are there in an avdp. pound?

   c. Which system are we using when we measure body weight by standing on a scale?

   d. Look in a metropolitan newspaper to find out how much money the silver you produced would bring on the open market. (1.00 troy oz. = 31.1 g)

Appendix A

Brief Bibliography of Resources

General Herbals

Lust, John B. *The Herb Book.* New York: Bantam Books, 1974.

Magic and Medicine of Plants. Pleasantville, NY: The Reader's Digest Association, Inc., 1986.

Rose, Jeanne. *Herbs and Things: Jeanne Rose's Herbal.* New York: Grosset & Dunlap, Workman Publishing Company, 1972.

Industrial Processes (including commercial dyes, metal processing, and soap making)

Kirk-Othmer Encyclopedia of Chemical Technology, 4th Edition. New York: John Wiley & Sons, Inc., 1991. [The 3rd edition of the *Encyclopedia* (1978) has an excellent article and bibliography on natural dyes.]

Medicinal Herbs (see also General Herbals)

Aronson, J.K. *An Account of the Foxglove anc Its Medical Uses, 1785-1985.* Oxford: Oxford University Press, 1985.

Duke, James A. *CRC Handbook of Medicinal Herbs.* Boca Raton: CRC Press, 1985.

Griggs, Barbara. *Green Pharmacy: A History of Herbal Medicine.* London: Robert Hale Ltd., 1981.

Sotheeswaran, S. "Herbal medicine: The scientific evidence," *Journal of Chemical Education,* 69, 1992, 444-446.

Weiner, Michael A. *The People's Herbal. A Family Guide to Herbal Home Remedies.* New York: Putnam Publishing Group, Perigee Books, 1984.

Wijesekera, R. O. B. *The Medicinal Plant Industry.* Boca Raton: CRC Press, 1991.

Natural Dyes

Casselman, Karen Leigh. *Craft of the Dyer: Colour from Plants and Lichens of the Northeast.* Toronto: University of Toronto Press, 1980.

Grae, Ida. *Nature's Colors: Dyes from Plants.* New York: Macmillan Publishing Company, 1974.

Kirk-Othmer Encyclopedia of Chemical Technology (see above under *Industrial Processes*).

Krochmal, Arnold, and Connie Krochmal. *The Complete Illustrated Book of Dyes from Natural Sources.* Garden City: Doubleday & Company, Inc., 1974.

Mayer, Fritz. *Chemistry of Natural Coloring Matters.* American Chemical Society Monograph 89, translated and revised by A. H. Cook. New York: Reinhold, 1943.

Oil Spills

Reader's Guide to Periodical Literature 1989. There are numerous articles listed under "Exxon Valdez." For follow-up and studies of environmental damage, check the 1990 and 1991 *Guide*, as well.

Flaherty, Michael L., ed. *Oil Dispersants: New Ecological Approaches.* ASTM Special Technical Publications. Philadelphia: ASTM, 1989.

Polymers and Non-Newtonian Fluids

Casassa, E.Z., A.M. Sarquis, and C.H. VanDyke. "The gelation of polyvinyl alcohol with borax," *Journal of Chemical Education*, 63, 1986, 57-60.

Walker, Jearl. "Serious fun with Polyox, Silly Putty, Slime and other non-Newtonian fluids," *Scientific American*, 239, Nov. 1978, 186-196.

Walker, Jearl. "Why do honey and syrup form a coil when they are poured?" *Scientific American*, 245, Sept. 1981, 216-224.

Soap

Mohr, Merilyn. *The Art of Soap Making: A Complete Introduction to the History and Craft of Fine Soapmaking.* Camden East, Ont.: Camden House Publishers, 1979.

Textbooks Which May Be Helpful

Brady, James E. *General Chemistry: Principles and Structure.* New York: John Wiley & Sons, 1990.

Brown, Theodore L., and H. Eugene LeMay, Jr. *Chemistry: The Central Science*, 4th Edition. Englewood Cliffs: Prentice Hall, 1988.

Chang, Raymond. *Chemistry*, 3rd Edition. New York: Random House, 1988.

Petrucci, Ralph H. *General Chemistry: Principles and Modern Applications*, 5th Edition. New York: Macmillan Publishing Company, 1989.

Appendix B

Chemicals Used in *Chemistry: The Experience*

This list is provided to give the student and instructor a quick overview of the laboratory chemicals used in this manual. Detailed instructions for solutions and lists of household chemicals and other common materials used are provided in the *Instructor's Manual for Chemistry: The Experience.*

Name	Formula
acetic acid	CH_3COOH
acetic anhydride	$C_4H_6O_3$
acetone	$(CH_3)_2CO$
adipoyl chloride	$ClOC(CH_2)_4COCl$
alizarin	$C_{14}H_8O_4$
aluminum	Al
aluminum sulfate	$Al_2(SO_4)_3$
ammonia, aqueous	$NH_3(aq)$
ammonium chloride	NH_4Cl
ammonium nitrate	NH_4NO_3
ammonium sulfate	NH_4SO_4
benzoic acid	C_6H_5COOH
bromothymol blue	$C_{27}H_{28}Br_2O_5S$
calcium	Ca
calcium carbonate	$CaCO_3$
calcium chloride	$CaCl_2$
calcium hydroxide	$Ca(OH)_2$
calcium nitrate	$Ca(NO_3)_2$
calcium oxide	CaO
calcium sulfate	$CaSO_4$
copper	Cu
copper(II) carbonate	$CuCO_3$
copper(II) chloride	$CuCl_2$
copper(II) nitrate	$Cu(NO_3)_2$
copper(II) oxide	CuO
copper(II) sulfate	$CuSO_4$
copper(II) sulfate pentahydrate	$CuSO_4 \cdot 5H_2O$
ethanol	C_2H_5OH
glucose	$C_6H_{12}O_6$
hexamethylenediamine	$H_2N(CH_2)_6NH_2$
hexane	C_6H_{14}
hydrochloric acid	HCl
hydrogen peroxide	H_2O_2
iodine solution	I_2 in KI

Chemicals, cont'd

Name	Formula
iron(III) nitrate	$Fe(NO_3)_3$
iron(III) oxide	Fe_2O_3
iron	Fe
isopropyl alcohol	C_3H_7OH
litmus paper, blue and red	
magnesium	Mg
manganese dioxide	MnO_2
methanol	CH_3OH
methyl red	$C_{15}H_{15}O_2N_3$
nitric acid	HNO_3
phenolphthalein	$C_{20}H_{14}O_4$
poly(vinyl alcohol), PVA	$(-CH_2CHOH-)$
potassium bromide	KBr
potassium carbonate	K_2CO_3
potassium chloride	KCl
potassium hydrogen phthalate	$KHC_8H_4O_4$
potassium hydrogen tartrate	$KHC_4H_4O_6$
potassium hydroxide	KOH
potassium iodate	KIO_3
potassium iodide	KI
potassium nitrate	KNO_3
potassium sulfate	K_2SO_4
potassium thiocyanate	$KSCN$
propanol	C_3H_7OH
salicylic acid	$C_7H_6O_3$
sand	SiO_2
silver chloride	$AgCl$
silver nitrate	$AgNO_3$
sodium borate	$Na_2B_4O_7 \cdot 10H_2O$
sodium bromide	$NaBr$
sodium carbonate	Na_2CO_3
sodium chloride	$NaCl$
sodium hydrogen carbonate	$NaHCO_3$
sodium hydrogen sulfite	$NaHSO_3$
sodium hydroxide	$NaOH$
sodium iodide	NaI
sodium metal (instructor only)	Na
sodium nitrate	$NaNO_3$
sodium phosphate	Na_3PO_4
sodium sulfate	Na_2SO_4
starch, soluble	
sucrose	$C_{12}H_{22}O_{11}$
sulfuric acid	H_2SO_4
urea	NH_2CONH_2
water	H_2O
zinc	Zn

Appendix C

Brief Introduction to Tie-dyeing

Tie-dyeing can be as simple or as complex as you wish to make it. Below are several figures and brief descriptions of the simplest folds and ties. The most important thing to remember is to tie the string very tightly to get maximum dye exclusion. Because at times you will be mixing dyes as the colors run together, and since dyes are chemicals undergoing reaction with the cloth and sometimes with each other, the colors produced when dyes run together may not be what you expected. Always experiment first. In your tie-dyeing experiments use 100% cotton cloth or T-shirts. Inexpensive remnant cloth can often be purchased wherever fabric is sold.

Some Beginning Steps

> These steps are necessarily brief. For detailed instructions, refer to the instructions with commercial dye or to one or more of the references in Appendix A.

1. A new T-shirt or handkerchief needs to be washed first to remove the sizing. If you are using a commercial dye, follow the instructions in the box. If you are experimenting with natural dyes on a T-shirt or large cloth, a handful of sodium carbonate (washing soda), $Na_2CO_3 \cdot 10H_2O$, in hot water will do. Rinse well.

2. The next step is to decide what the background color of the cloth will be. You can leave a T-shirt white or dye it a light color. You might wish to dye the sleeves one color and the body another. There aren't any rules, except that it will be easier to dye over a light color, so keep that in mind when you pick the background color.

3. Apply the background color, using the instructions with the commercial dye or one of the references in Appendix A for help.

4. Experiment with ties, using the ones below and others you make up.

5. If you wish to dye only one tied area, dip the tied area in the boiling dye solution for a few moments. This will take some experimentation, so good note-taking will help. Another tie can be dipped in another dye solution. When you have finished, remove the ties and rinse the cloth well. Wash the cloth with detergent and dry.

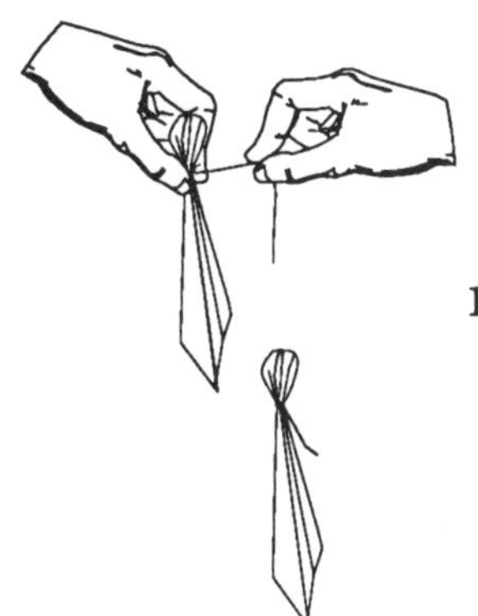

Some Ties

1. Pinch a small bunch of cloth with the fingers of one hand and tie string around the bundle with the other. The result is a circle of background color where the string was tied with the new color on the exposed area. This gives a sunburst effect.

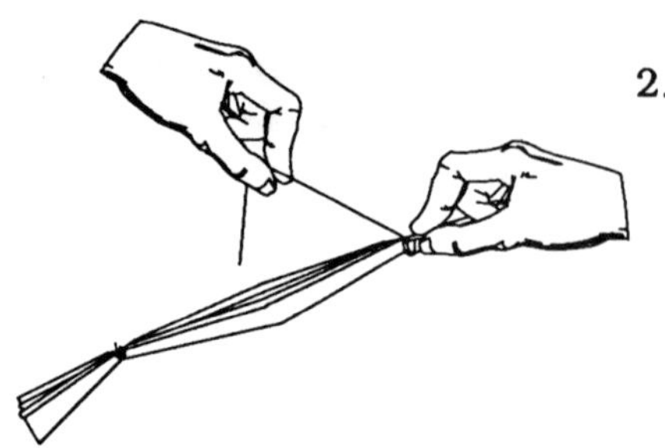

2. Fan-fold a small portion of the cloth with narrow folds. Tie string along the length of the fold in numerous places by first tying the ends of the fold, as shown. After the ends are tied, make ties along the fold at varying intervals. The result will be long bursts of color with tie marks across them.

3. Fan-fold again, but this time make the folds fairly wide. Roll the folded material very tightly, and then tie string tightly around the ends to keep the roll together and to exclude dye from the middle of the folds. Dip one end of the rolled material in one dye and the other end in another color, leaving the middle the background color.

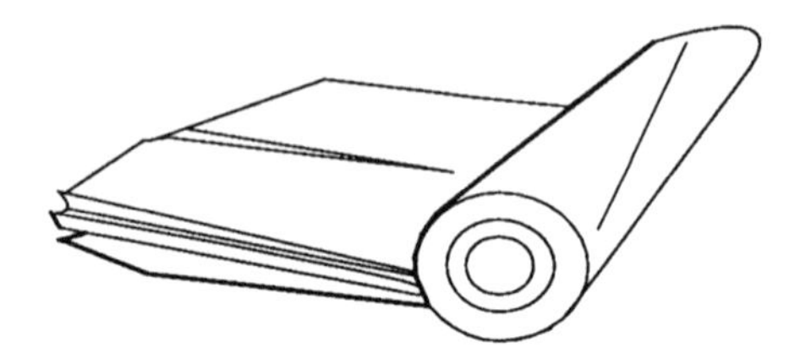

4. Draw some ideas for your own folds here.

Appendix D

More Dyes

If you did Inquiry 24 you probably experimented with onion skin and quercetin, the dye that is extracted from it. There are many other important natural dyes, and the references in Appendix A will give you detailed information about them. You may find the low toxicity dye *alizarin* in your chemistry storeroom, and you will find *turmeric* on the grocery spice shelf.

Alizarin

Alizarin is extracted from the roots of a plant called madder, a member of the family Rubiaceae which grows in Europe and Asia. It is one of the oldest dyes, known since antiquity, and the only natural dye used to produce a good strong red known as Turkey red. It can be mordanted with alum, or with alum and lime (CaO). It was first isolated and identified in 1826 and since that time has been synthesized commercially. The alizarin you get in the chemistry storeroom is most likely synthetic and therefore much less expensive than dye which has been extracted from roots of madder plants. Its chemical name is 1,2-dihydroxyanthraquinone.

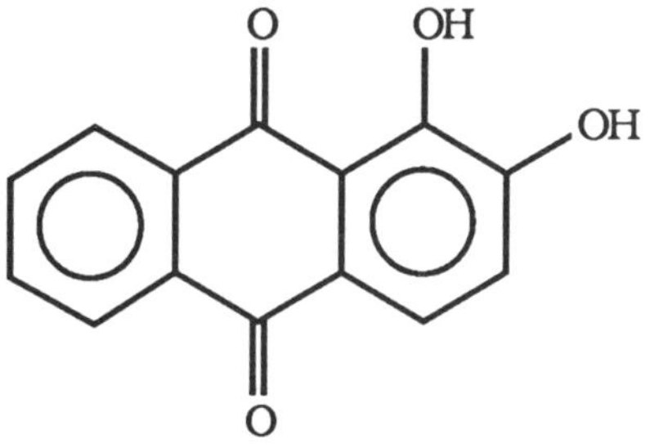

Alizarin

Turmeric

The pigment in the spice turmeric is curcumin, a yellow-orange dye extracted from the roots of plants of the genus *Curcuma* (family Zingiberaceae) which grows principally in Asia and the East Indies. It is related to saffron, the expensive yellow substance used as a spice and for yellow color in food in Asia, and, like saffron, it is a carotenoid. (See Inquiry 22.) It dyes cotton yellow-orange without a mordant, and if vinegar is added to the dye solution the result is a bright yellow.

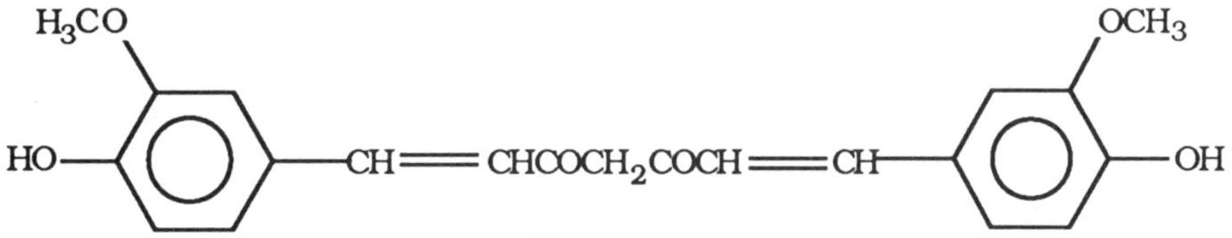

Curcumin